Source Code

the balance of persistence

Thad Roberts

Source Code: the balance of persistence

By Thad Roberts

First Edition

Print Date 03-18-2021

Published in the United States by Thad Roberts

ISBN 978-0-9963942-6-0

Cover by Pashaa Sanwick
Figures by Thad Roberts

Other books by Thad:

Einstein's Intuition: Visualizing Nature in Eleven Dimensions

Moon Rock: Mare Crisium

Passages

A Perfect Universe

44

To Anaximander and all those that notice.

Preface — the search

> *"The task is not to see what has never been seen before, but to think what has never been thought before about what you see every day."*
>
> *Erwin Schrödinger*

In an early attempt to construct a story that accounts for reality's persistent physical properties, Plato advanced the idea that the elemental building blocks of geometry are also the elemental building blocks of reality. That is, Plato proposed that the 5 simple shapes that are constructible from faces whose sides are all of equal length (the Platonic solids) are *atoms* of fire, air, earth, water, and aether—the primordial substance.

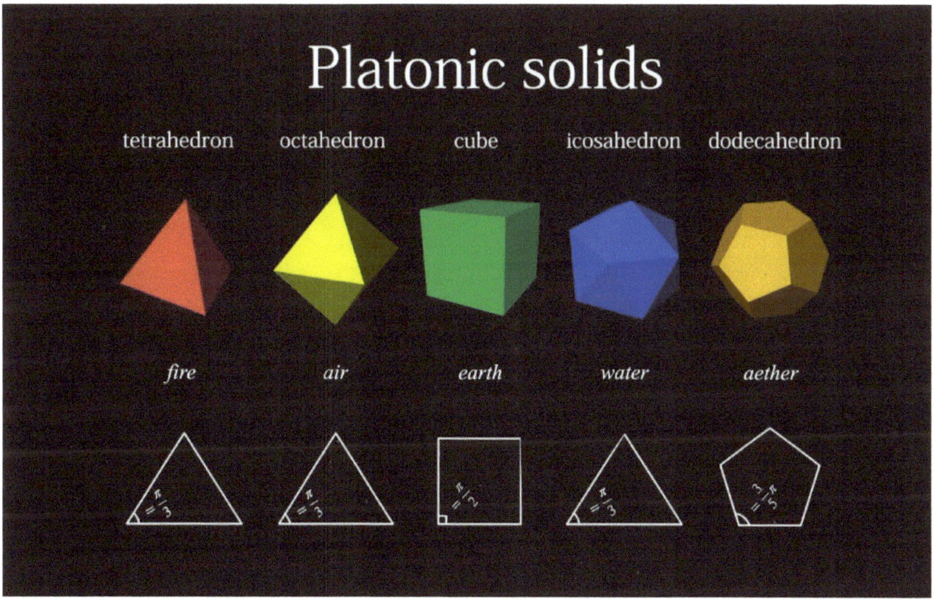

The 5 Platonic solids

If *geometry* is fundamentally responsible for the constructs of physical reality, then humanity has a path towards implicitly obtaining the logic of existence. With the light of Plato's 'candle' it becomes possible to walk ourselves out of the cave of ignorance.

For eons, answering the call of existence, and making one's best attempt to make sense of things, meant studying the Platonic solids. If you wanted to better understand reality, you would spend your hours staring at these shapes, trying to divine anything else these elemental constructs might have to say, trying to figure out how they combine to build the world we see. And yet, despite all the people examining these ideal mathematical objects for further clues, despite the specific effort to understand them as well as possible, a basic connection between these geometric shapes went unnoticed for 2000 years.

When Leonard Euler looked at the Platonic solids, he noticed that they are connected by a single elementary geometric relationship. Every one of them maintains the same balance of zero, one, and two-dimensional features (vertices, edges, faces). That is, each shape's number of vertices, minus its number of edges, plus its number of faces is always equal to 2.

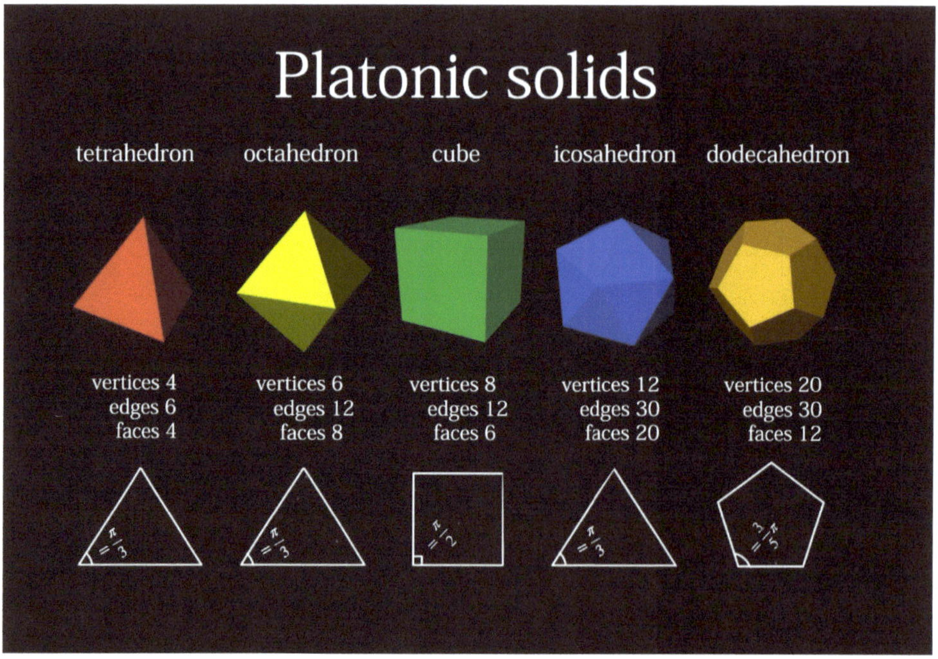

Euler's surprise relationship

$$\text{vertices} - \text{edges} + \text{faces} = 2$$

$$V - E + F = 2$$

However obvious and simple this fact may now appear, the question that we must address is, "How did everyone else miss it?" Despite intense interest in these objects, and the generations of people searching them for clues, nobody had ever found what Euler found. Why? It doesn't take much time to count up these properties and compare them. Yet nobody had. Why not?

The answer is that, before Euler, nobody had imagined that an object's *number of sides* was a feature worthy of attention. They had never seen anyone else care about such a feature, so it just hadn't occurred. Euler was the first to count up these features and compare them because he was the first to imagine the option.

Conscious thinkers have only two routes to new thought. Through example, or through genuine new reaches of imagination. Just as it is usually easier to verify an answer than to find one, it is usually easier to follow a line of logic laid out for you, than to lay one out for yourself. Which is why conscious thinkers almost always follow.

The trouble is, that if we are always just following, then we can get seriously stuck. If nobody has ever shown us an example of being interested in the number of edges an object has, we can all stare at something for 2000 years without seeing what's right in front of our eyes.

Euler looked with new eyes, under his own investigation, instead of looking as he had been directed to. And in that moment the world inherited new conceptual powers.

Another famous example of how difficult it can be to take the next logical next step without anyone showing us how, comes from our history of conceptualizing gravity. For millennia humans personally pondered *why* things fall to the ground, but it wasn't until Galileo came along that anyone thought to measure *how* things fall under the influence of gravity.

Galileo's measurements could be performed by nearly any human within minutes, with no cost (like rolling a ball down an inclined plane and marking the position of the ball down the ramp at successive increments of time and then comparing the distance between those marks). But before Galileo nobody had tried. Nobody had thought to.

By extending the powers of *measurement* to gravity Galileo became the first to see it more clearly. What he found is that gravity's action depends on an inverse square law. In other words, he discovered that gravity is geometric.

This fact about gravity had always been true, it had always been staring us in the face, screaming from every freefall occurrence ever witnessed, but nobody had seen it before. Because nobody thought of gravity as a thing to physically measure, a thing to look into that way.

Through his efforts Galileo came to one of the greatest insights of all time, that the truth about reality is right in front of our eyes. The task is simply to see more clearly, to more explicitly conceive existence.

"Philosophy is written in this grand book, the universe, which stands continually open to our gaze. But the book cannot be understood unless one first learns to comprehend the language and read the letters in which it is composed. It is written in the language of mathematics, and its characters are triangles, circles, and other geometric figures without which it is humanly impossible to understand a single word of it; without these, one wanders about in a dark labyrinth."

Galileo Galilei

Today's best geometric description of the fundamental structure of reality is called *quantum field theory*. This construction naturally produces *all* of the actions of quantum mechanics and special relativity, a feat that earns it the title of being the pinnacle achievement of science. The problem is that although this construction makes the right predictions, nobody understands why it has that particular construction. The parameters of quantum field theory have no story themselves. Every single one of them remains utterly unexplained, known from experimental measurement only.

That is, we cannot explain *why* quantum field theory is constructed as it is, we cannot predict its parameters to any degree of accuracy at all, but once we construct our field theory with those parameters the actions of quantum mechanics and special relativity are reproduced in full.

Today, any serious investigation of reality involves staring at the parameters of quantum field theory, in search of the logic connecting them. That is, the modern version of the quest to tell the accurate story of reality's construction (now called the theory of everything) quite literally boils down to explaining the parameters of quantum field theory, explaining why they are what they are.

the parameters of quantum field theory

$m_e = 9.1093837015(28) \times 10^{-31}\ kg$	electron mass
$m_+ = 1.67262192369(51) \times 10^{-27}\ kg$	proton mass
$m_N = 1.67492749804(95) \times 10^{-27}\ kg$	neutron mass
$m_c = 2.272(63) \times 10^{-27}\ kg$	charm quark mass
$m_d = 8.3(07) \times 10^{-30}\ kg$	down quark mass
$m_u = 3.92(89) \times 10^{-30}\ kg$	up quark mass
$m_s = 1.69(16) \times 10^{-28}\ kg$	strange quark mass
$m_b = 7.45(07) \times 10^{-27}\ kg$	beauty (bottom) quark mass
$m_t = 3.084(07) \times 10^{-25}\ kg$	truth (top) quark mass
$m_H = 2.2315(28) \times 10^{-25}\ kg$	Higgs boson mass
$m_Z = 1.625566(38) \times 10^{-25}\ kg$	Z boson mass
$m_W = 1.43288(21) \times 10^{-25}\ kg$	W boson mass
$m_\tau = 3.16754(21) \times 10^{-27}\ kg$	tau mass
$m_\mu = 1.883531627(42) \times 10^{-28}\ kg$	muon mass
$m_{\nu_\tau} => 0\ kg\ ???$	tau neutrino mass
$m_{\nu_\mu} => 0\ kg\ ???$	muon neutrino mass
$m_{\nu_e} => 0\ kg\ ???$	electron neutrino mass

$\alpha = 7.2973525698(24) \times 10^{-3}$	fine-structure constant
$e = 1.602176565(35) \times 10^{-19}\ C$	electron charge
$\lambda_C = 2.4263102389(16) \times 10^{-12}\ m$	Compton wavelength
$K_J = 4.835978484 \times 10^{14}\ sC/m^2kg$	Josephson constant
$\hbar = 1.054571726(47) \times 10^{-34}\ m^2kg/s$	Planck's constant
$\varepsilon_0 = 8.8541878128(13) \times 10^{-12}\ s^2C^2/m^3kg$	electric constant
$\kappa = 8.9875517923(14) \times 10^{9}\ m^3kg/s^2C^2$	Coulomb's constant
$H_C = 3.87404614(17) \times 10^{-5}\ C^2/m^2kg$	quantized Hall conductance
$\mu_B = 9.274009994(57) \times 10^{-24}\ m^2C/s$	Bohr magneton
$\mu_0 = 1.256637061 \ldots \times 10^{-6}\ mkg/C^2$	magnetic constant
$c_1 = 3.741771852 \ldots \times 10^{-16}\ m^4kg/s^3$	1st radiation constant
$\sigma_e = 6.6524616(18) \times 10^{-29}\ m^2$	electron Thomson cross section
$G_0 = 7.748091729 \ldots \times 10^{-5}\ sC^2/m^2kg$	conductance quantum
$a_0 = 5.2917721092(17) \times 10^{-11}\ m$	Bohr electron radius
$m_u = 1.66053906660(50) \times 10^{-27}\ kg$	atomic mass constant
$Z_0 = 3.76730313668(57) \times 10^{2}\ m^2kg/sC^2$	characteristic impedance
$\sigma = 5.670374419 \times 10^{-8}\ kg/s^3K^4$	Stefan-Boltzmann constant
$N_A = 6.02214076 \times 10^{23}\ 1/mol$	Avogadro constant
$R_K = 2.58128074434(84) \times 10^{4}\ m^2kg/sC^2$	von Klitzing constant
$E_h = 4.3597447222071(85) \times 10^{-18}\ m^2kg/s^2$	Hartree energy
$c = 2.99792458 \times 10^{8}\ m/s$	speed of light

$c_{1L} = 1.191042869(53) \times 10^{-16}\ m^4kg/s^3$ spectral radiance
$r_e = 2.8179403227(19) \times 10^{-15}\ m$ classical electron radius
$\mu_N = 5.050783699(31) \times 10^{-27}\ m^2C/s$ Nuclear magneton
$c_2 = 1.438776877 \ldots \times 10^{-2}\ m\ K$ 2nd radiation constant
$g_\mu = -2.0023318418(13)$ muon g-factor
$g_e = -2.00231930436256(35)$ electron g-factor
$R = 8.314462618\ m^2kg/s^2K\ mol$ molar gas constant
$\Phi_0 = 2.067833848 \ldots \times 10^{-15}\ m^2kg/sC$ magnetic flux constant
$q_c = 3.6369475516(11) \times 10^{-4}\ m^2/s$ quantum of circulation
$g_+ = +5.5856946893(16)$ proton g-factor
$R_\infty = 1.0973731568539(55) \times 10^7\ 1/m$ Rydberg constant
$\gamma_+ = 2.6752218744(11) \times 10^8\ s/kg\ C$ proton gyromagnetic ratio
$N_\mu = -9.6623647(23) \times 10^{-27} C\ m^2/s$ neutron magnetic moment
$F = 9.648533212 \ldots \times 10^4\ C/mol$ Faraday constant
$g_N = -3.82608545(90)$ neutron g-factor
$\alpha_G = 1.7518(21) \times 10^{-45}$ gravitational coupling constant
$\omega_c = 7.763441 \times 10^{20}\ 1/s$ Compton angular frequency
$S_{mi} = 4.419 \times 10^9\ kg/sC$ Schwinger magnetic induction
$G = 6.67384(80) \times 10^{-11}\ m^3/s^2kg$ gravitational constant
$k_B = 1.380649 \times 10^{-23}\ m^2kg/s^2K$ Boltzmann constant
$r_+ = 8.751(61) \times 10^{-16}\ m$ proton radius

Where the digits in the parentheses define the measurement error in the preceding two digits (e.g. $8.751(61)$ means 8.751 ± 0.061), and the neutrino masses are only known to be non-zero.

This book tells the story of these numbers, revealing that the combinatorial logic behind the structure of reality is mapped by the minimal self-balanced manifold—the hyperbolic figure eight knot. That is, in this book we notice for the first time that the ideal minimum geometry, the one that defines the minimal possible persistent stage, is the stage of physical reality.

To define the unique geometric properties of this minimally persistent manifold, we fully decompose the hyperbolic figure eight knot under ideal partition balance (Chapter 0-4), and discover (Chapter 5) that the boundary conditions of this decomposition precisely predict the Planck constants (the mysterious limits of time, space, charge, mass, and temperature—which have also only been experimentally known until now). Then we find that the distinct geometric characteristics of this minimal stage (the unique actions of its partition balance) perfectly prescribe the parameters of quantum field theory—the entire set (Chapter 6-10).

In summary, in this book, we discover that the set of fundamental structural traits of physical reality (the elemental partitions of *charge* and *mass*, and the *constants of Nature*) are the set of unique geometric properties that define the simplest manifold balance—the hyperbolic figure eight knot.

Chapter 0 — the simplest manifold

In the mid 1970's Robert Riley and Troels Jorgensen independently discovered that the figure eight knot emits a complement with hyperbolic structure. This was the first known example of a hyperbolic knot—a closed loop under hyperbolic balance.

The complement of the figure-eight knot (portrayed here with cut-outs to allow visibility) is the finite volume whose simple closed geodesics trace out the figure-eight knot.

The hyperbolic figure eight knot is a double cover of the Gieseking's manifold, the simplest among all the non-compact hyperbolic 3-manifolds. As such, it defines the minimally self-balanced stage. In other words, *persistence* is minimally obtained when a filament (a finite wave element) becomes balanced under local cyclic action, twisting and folding into the shape of the hyperbolic figure eight knot.

The internal volume complement of this minimal geometry is equal to twice Gieseking's constant (G_{Gi}).

$$2G_{Gi} = V_{fe} = 2.02988321281930\ldots$$

Intrigued by this ideal geometry, William Thurston looked closer and discovered that the complement of the figure-eight knot decomposes into a union of two regular ideal hyperbolic tetrahedra. (He then famously formulated his geometrization conjecture, claiming that all 3-manifolds admit a certain kind of geometric decomposition involving 8 geometries, most of which are hyperbolic. Grigori Perelman proved this conjecture in 2002-2003.)

Thurston's decomposition was a brilliant addition to our knowledge of the hyperbolic figure eight knot's geometry, but it didn't tell us everything there is to know. That is, it wasn't explicit enough to define every feature of this minimal geometric balance.

For a complete understanding of the hyperbolic figure eight knot's geometric properties (to sharply define its boundary conditions and every unique property of its balance) we need to go from generally decomposing its internal complement volume to explicitly decomposing its complete partition balance.

Chapter 1 — the minimum partition balance

The hyperbolic figure eight knot decomposes into 5 unique geometrically bound balances.

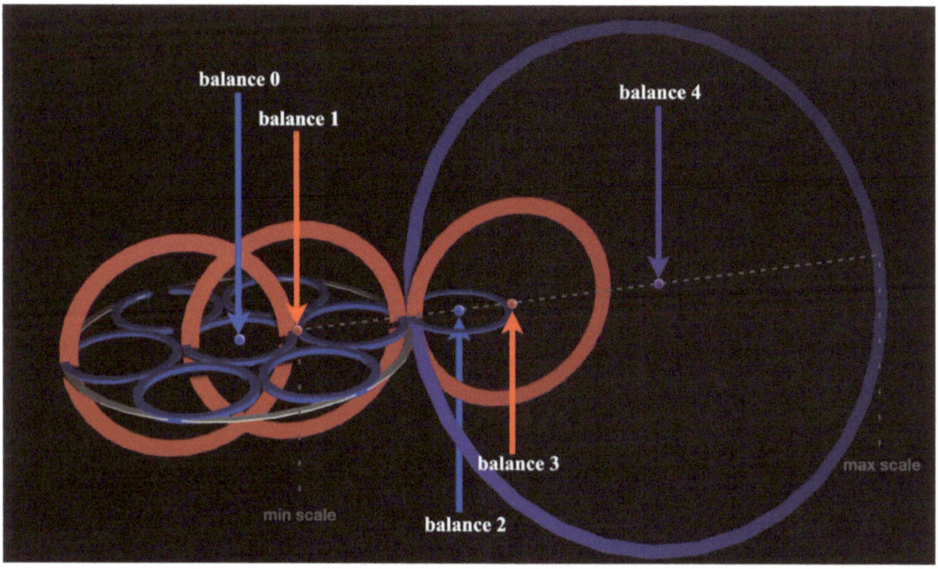

The partition balance of the hyperbolic figure eight·knot.

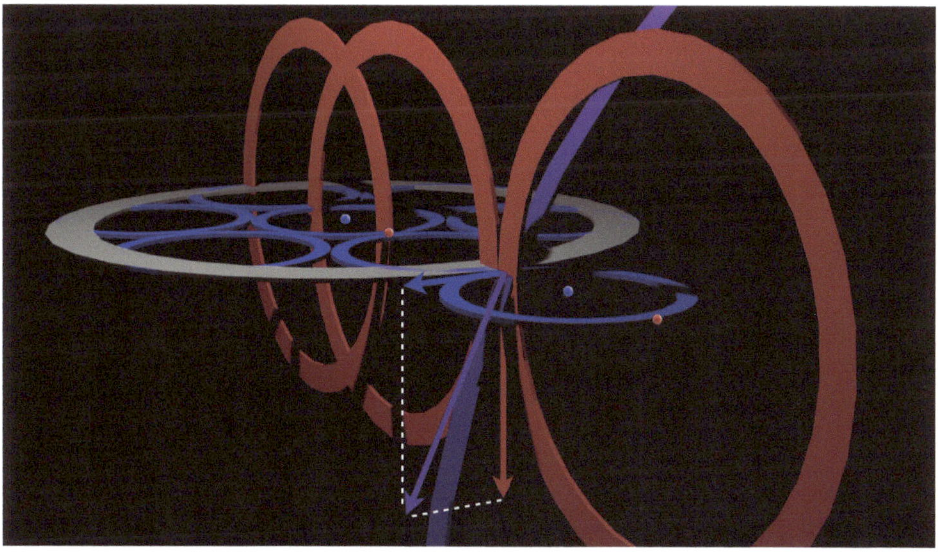

The combined action of balance 2 and 3 (blue and red arrows) rotates the external domain under balance 4 (purple).

general decomposition map

$$(\text{inner action})_4{}^{-1}(\text{outer rotation})_4{}^{-1} = (\#\text{of hinges})_4{}^{-1} \quad (\text{boundary } 4)^{-1}$$

$$(\text{the external reference}) = 1 \quad \text{balance } 4$$

$$(\text{inner action})_3(\text{outer rotation})_3 = (\#\text{of hinges})_3 \quad \text{balance } 3$$

$$(\text{inner action})_2(\text{outer rotation})_2 = (\#\text{of hinges})_2 \quad \text{balance } 2$$

$$(\text{inner action})_4(\text{outer rotation})_4 = (\#\text{of hinges})_4 \quad \text{boundary } 4$$

$$(\text{inner action})_1(\text{outer rotation})_1 = (\#\text{of hinges})_1 \quad \text{balance } 1$$

$$(\text{inner action})_0(\text{outer rotation})_0 = (\#\text{of hinges})_0 \quad \text{balance } 0$$

Where the union of balance 0 and balance 1 defines the internal double-cover domain, balance 4 (created by the union of balance 2 and balance 3) defines the external split-cover domain, and "hinges" define *vertices of action*.

The balance point of the external domain (balance 4) prescribes the unitary scale of this decomposition (1)—the scale of stability projected under this action. Under this ideal inversion, the *boundaries* of balance 4 keep balance with the other 4 balances.

The action of the hyperbolic figure eight knot is localized, therefore the rotations of its internal and external parts perfectly balance. Since the internal partitions of that balance are arranged under ideal *split* factorization $\sqrt{!n - b(n)}$, the external partitions must be arranged as squared and *doubled* expressions of those internal actions. And since there are 3 external actions balancing against 2 internal actions in this geometry, the external expressions of that arrangement are its unique phase expressions $(-1,0,1)$.

external

$$(\#\text{of hinges})_3 = 2\left(\sqrt{!n - b(n)} - 1\right)^2 = 2(3-1)^2 = 8$$

$$(\#\text{of hinges})_2 = 2\left(\sqrt{!n - b(n)} \pm 0\right)^2 = 2(3 \pm 0)^2 = 18$$

$$(\#\text{of hinges})_4 = 2\left(\sqrt{!n - b(n)} + 1\right)^2 = 2(3+1)^2 = 32 = 2^n$$

internal

$$(\#\text{of hinges})_1 = b(n) = 35$$

$$(\#\text{of hinges})_0 = !n = 44$$

This localization symmetry updates the hyperbolic figure eight knot's decomposition map to:

decomposition map

$(\text{inner action})_4{}^{-1} e^{-\phi_4} = 2^{-n}$	$(\text{boundary } 4)^{-1}$
$(\text{external reference}) = 1$	balance 4
$(\text{inner action})_3\, e^{\phi_3} = 2(3-1)^2$	balance 3
$(\text{inner action})_2\, e^{\phi_2} = 2(3 \pm 0)^2$	balance 2
$(\text{inner action})_4\, e^{\phi_4} = 2(3+1)^2 = 2^n$	boundary 4
$(\text{inner action})_1\, e^{\phi_1} = b(n)$	balance 1
$(\text{inner action})_0\, e^{\phi_0} = !n$	balance 0

Where ϕ_0, ϕ_1, ϕ_2, ϕ_3, & ϕ_4 are the magnitudes of external rotation keeping each balance $(\text{outer rotation})_k = e^{\phi_k}$, the union of balance 0 and balance 1 defines the internal double-cover domain, balance 4 (created by the union of balance 2 and balance 3) defines the external split-cover domain, $n = 5$, the number of unique rotations partitioning the balance of the hyperbolic figure eight knot, $!n = 44$, the number of derangements available to those 5 rotations, and b = the break in scale symmetry between the first and second balances maintained under this action (see Chapter 3).

To complete this decomposition, we must specify its balance of inner actions. Let's begin by defining the absolute minimum cyclic expression (balance 0).

Chapter 2 — the minimum limit of persistence

The minimum cyclic action of the hyperbolic figure eight knot defines the minimum *possible* cyclic action, a circularly closed action πr^2 trivially balanced against an external counter-rotation e^{ϕ_0}.

$$\pi r^2 e^{\phi_0} = !n \qquad \text{balance 0}$$

Where $n = 5$ the number of unique rotations partitioning this balance, $!n = 44$ the number of derangements available to those 5 rotations, and $\phi_0 =$ the magnitude of the external rotation of this minimum cyclic action.

To represent this minimal action factoring into rotations that counterbalance under ideal complex division (splitting into equal but opposite internal rotations, each absorbing half the input), we set r equal to the hyperbolic sine function.

$$\frac{1}{2}(e^x - e^{-x}) = sinh(x) = r$$

And we set the argument of this action (x) equal to the minimal representation of square split division—a number that constructively possesses complex four-fold symmetry.

$$x = \left(\frac{1}{2}\right)^2 = \left(\frac{1}{-2}\right)^2 = -\left(\frac{1}{2i}\right)^2 = -\left(\frac{1}{-2i}\right)^2$$

$$\frac{1}{2}\left(e^{\left(\frac{1}{2}\right)^2} - e^{-\left(\frac{1}{2}\right)^2}\right) = sinh\left(\left(\frac{1}{2}\right)^2\right) = r$$

Plugging this ideally hyperbolically balanced r into our equation, we arrive at a precise characterization of the minimum cyclic limit of *persistence*—the smallest cyclic action maintained by the square split-division balance of the hyperbolic figure eight knot.

$$\pi\left(sinh\left(\left(\frac{1}{2}\right)^2\right)\right)^2 e^{\phi_0} = !n \qquad \text{balance 0}$$

Where $\phi_0 = 5.391258368\ldots$ the magnitude of the external rotation that balances this minimum cyclic action, $sinh(x) =$ the hyperbolic sine function, $n = 5$, the number of unique rotations partitioning this balance, and $!5 = 44$, the number of derangements available to those rotations.

This equation defines the minimum partitioning that can be balanced under locally-closed cyclic action. Therefore, it holds the honor of defining the *ultimate* boundary condition, the absolute minimum limit of measure, beyond which the possibility for *persistence* itself is operationally cut off. In other words, this equation describes the minimum boundary of time.[1]

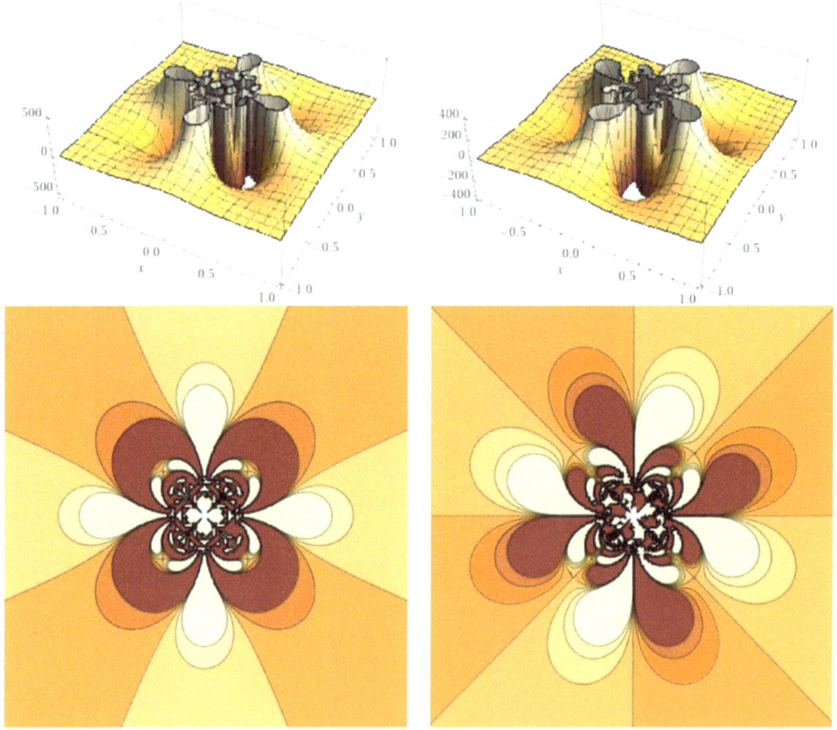

Real (left) and Imaginary (right) complex plots of the minimum partition balance maintained by the minimum (0th) *cyclic action of the hyperbolic figure eight knot.*

---

[1] Equivalent expressions for the minimum limit of persistence include:

$$\pi\left(i\,sin\left(\left(\frac{\sqrt{i}}{2}\right)^{2}\right)\right)^{2} e^{\phi_0} = !\,\boldsymbol{n} \qquad \text{balance } 0$$

and

$$\frac{\pi}{2}\left(\frac{1}{2}\left(e^{-1/2} + e^{1/2}\right) - 1\right) e^{\phi_0} = !\,\boldsymbol{n} \qquad \text{balance } 0$$

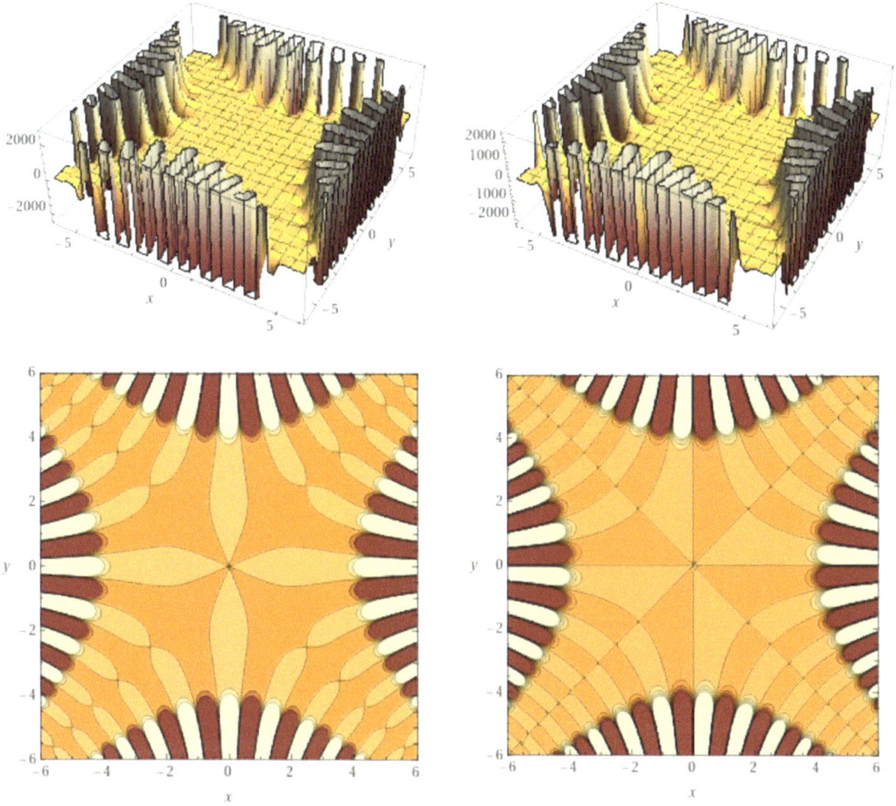

Real (left) and Imaginary (right) inverse complex plots of how the minimum balance partitions.

Chapter 3 — the first consequence

Under this construction, the highest resolution of cyclic action (the smallest scale) refers to the smallest closed loop (represented 2-dimesionally as a single circle). Zooming out to the next available scale (one step out in resolution) the next sized circle we can construct from those circles is comprised of 7 elemental circles (6 outer circles surrounding 1 inner circle) and the negative space that divides them. Therefore, the break in scale-symmetry between the first and second scales of circular construction is $\boldsymbol{b} = 7$.

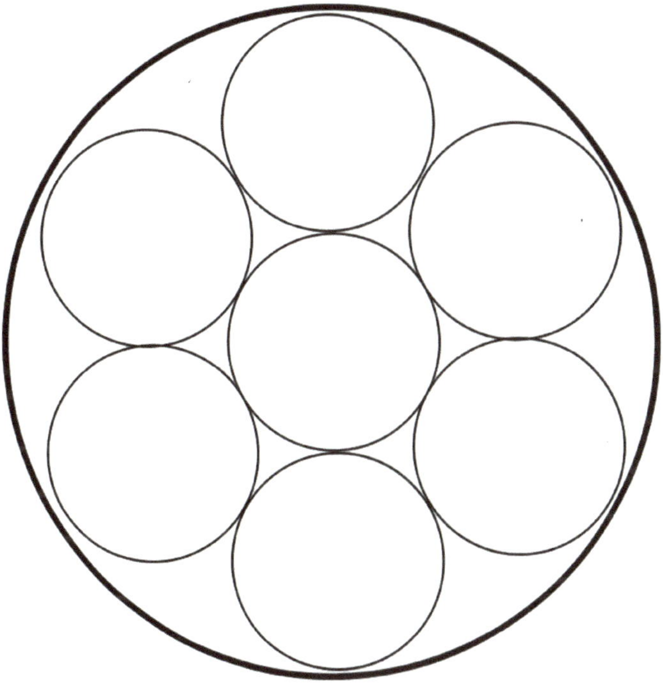

$$\frac{\text{\# of circles included on } 2^{\text{nd}} \text{ scale}}{\text{\# of circles included on } 1^{\text{st}} \text{ scale}} = \boldsymbol{b} = 7$$

With this in mind, let's zoom out to the next balance of the hyperbolic figure eight knot (balance 1), whose inner action is the operationally-rotated inverse action of balance 0 adjusted for the break in scale symmetry between the 2 balances. Balance 1 maintains a new external rotation (ϕ_1) that, from this scale on out, rotates in sync over ($\boldsymbol{b}(\boldsymbol{n}) = 35$) of the fundamental ($!\boldsymbol{n} = 44$) hinges.

$$\left(\sinh\left(\sinh\left(\frac{1}{b}\right)\right)\right)^{-1} e^{\phi_1} = b(n) \qquad \text{balance 1}$$

Where $\phi_1 = 1.616259181\ldots$ is the magnitude of the external rotation maintaining balance 1, $\sinh(x)$ = the hyperbolic sine function, b = the break in scale symmetry between the first and second scales of balanced action in the hyperbolic figure eight knot, and $b(n) = 35 =$ the number of hinges (derangements) this balance partitions over.[2]

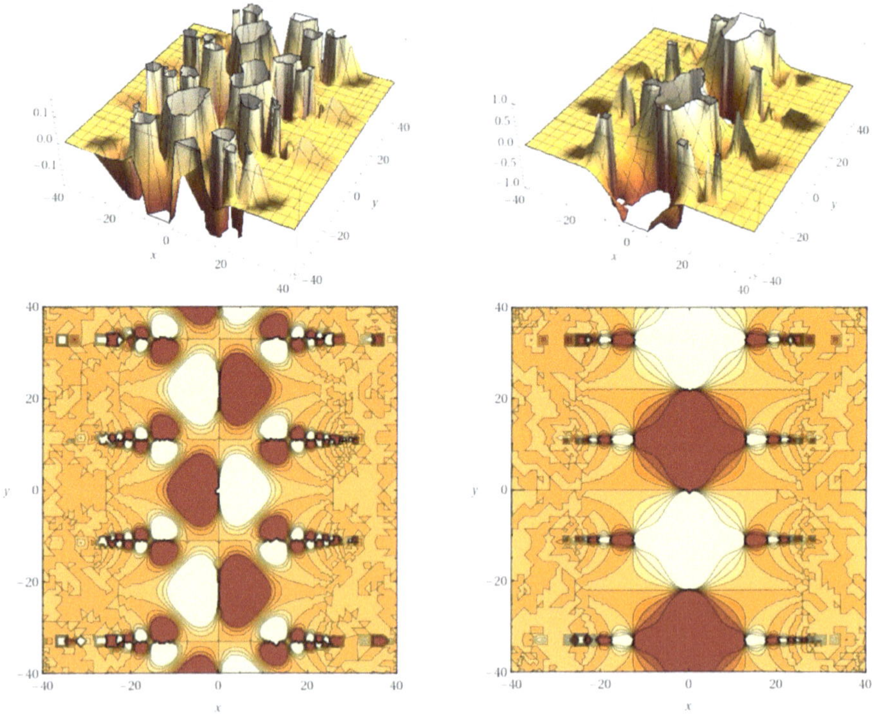

Real (left) and Imaginary (right) complex plots of the partition balance maintained by the 1st cyclic action of the hyperbolic figure eight knot.

---

[2] Balance 1 can be equivalently written as:

$$\left(\frac{1}{2}\left(e^{\left(\frac{1}{2}\left(e^{1/b} - e^{-1/b}\right)\right)} - e^{\left(\frac{1}{2}\left(e^{-1/b} - e^{1/b}\right)\right)}\right)\right)^{-1} e^{\phi_1} = b(n)$$

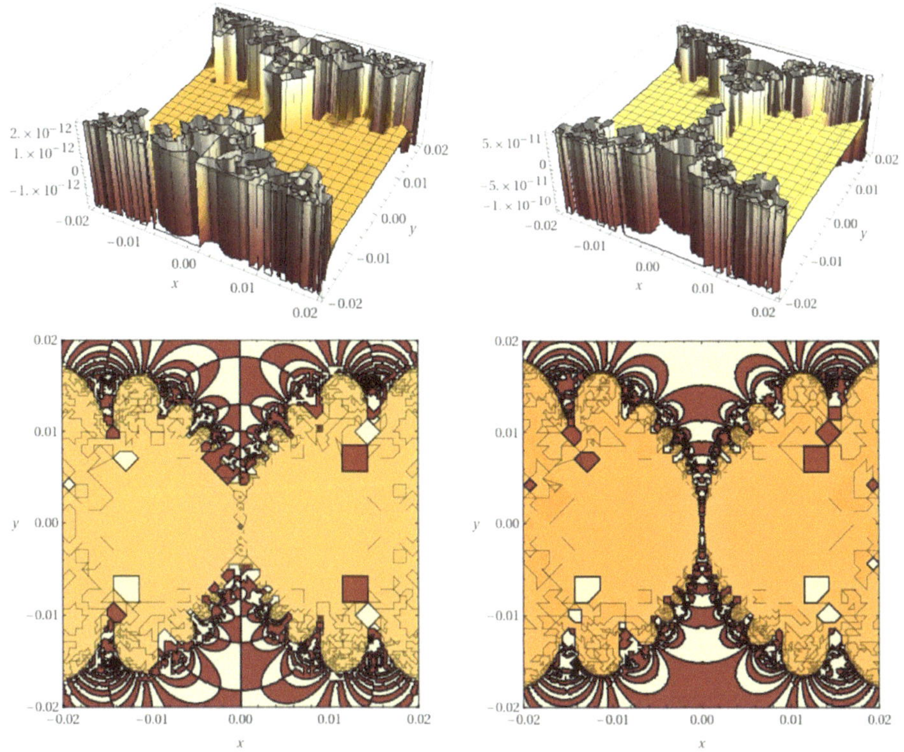

Real (left) and Imaginary (right) inverse complex plots of how the action of the 1st balance partitions.

Balance 0 and balance 1 trivially maintain an arrangement of squared partitions (n^2) that orthogonally factor under ideal hyperbolic balance ($\Gamma(n)$).

$$b \, cosh(log \, b) = n^2$$

$$b \, sinh(log \, b) = \Gamma(n)$$

Where $cosh(x) =$ the hyperbolic cosine function, $sinh(x) =$ the hyperbolic sine function, $log(x) =$ the logarithm function, $\Gamma(x) =$ the Gamma function which encodes hyperbolically balanced partitions, $n = 5$ the number of unique rotations partitioning the balance of the hyperbolic figure eight knot and, $b = 7$ the break in scale symmetry of this balance.

Chapter 4 — the external balances

Within the external domain of the hyperbolic figure eight knot the boundaries of balance 2 and balance 3 are put in mutual ideal hyperbolic vortex arrangement (see Chapter 6). Balance 2 forms the vortex's throat, defining where the n unique partitions of the hyperbolic figure eight knot doubly-periodically divide W_{We}, while factorizing along the break in scale symmetry of the geometry $\sqrt{b}$ and wrapping axisymmetrically around a single pole $(3)^{1/3}$.

$$\frac{n}{W_{We}\,\sqrt{b}\,(3)^{1/3}}e^{\phi_2} = 2(3 \pm 0)^2 \qquad \text{balance 2}$$

Where $\phi_2 = 1.875545967 \ldots$ is the magnitude of the external rotation that maintains this balance, $n =$ the number of unique rotations partitioning the balance of the hyperbolic figure eight knot, $2(3)^2 = 18 =$ the number of hinges (derangements) partitioning balance 2, and the Weierstrass constant is the unitary balance of the Weierstrass sigma function $W_{We} = \frac{1}{2}\sigma(1|1,i)$.

$$W_{We} = (\,2^{2n}\,e^{\pi}\,)^{1/8}\frac{\Gamma\left(\frac{1}{2}\right)}{\left(\Gamma\left(\left(\frac{1}{2}\right)^2\right)\right)^2}$$

Where $\Gamma(x) =$ the Gamma function, and $\Gamma\left(\frac{1}{2}\right) = \sqrt{\pi}$.

The unitary Weierstrass balance defines a harmonic double cover projection of the n-dimensional hypersphere, internally divided over eight partitions and maintained under lemniscate balance (divided by $2\sqrt{2}\,L$).

$$\mathrm{W_{We}} = (\,2^{2n}\,e^{\pi}\,)^{1/8}\,\frac{1}{2\sqrt{2}\,L}$$

Where $L =$ the lemniscate constant,

$$L = \frac{1}{2\sqrt{2}}\frac{\left(\Gamma\left(\left(\frac{1}{2}\right)^2\right)\right)^2}{\Gamma\left(\frac{1}{2}\right)}$$

and e^{π} = the infinite sum of even-dimensional (harmonic) volumes; which follows from the fact that the equations for the volume and surface areas of n-dimensional hyperspheres of radius r are:

$$V_n(r) = \frac{\pi^{n/2}}{\Gamma\left(\frac{n}{2}+1\right)} r^n \qquad S_{n-1}(r) = \frac{2\pi^{n/2}}{\Gamma\left(\frac{n}{2}\right)} r^{n-1}$$

$$\lim_{k\to\infty} \frac{\pi^0}{\Gamma(1)} + \frac{\pi^1}{\Gamma(2)} + \frac{\pi^2}{\Gamma(3)} + \cdots + \frac{\pi^k}{\Gamma(k+1)} = e^{\pi}$$

$$\lim_{k\to\infty} \frac{\pi^0}{0!} + \frac{\pi^1}{1!} + \frac{\pi^2}{2!} + \frac{\pi^2}{3!} + \cdots + \frac{\pi^k}{\mathrm{k}!} = e^{\pi}$$

Where, again, $\Gamma(x)$ = the Gamma function.

Alongside the balance of this unitary Weierstrass sigma function, its associated Weierstrass elliptic function orthogonally maintains a balance with equianharmonic, lemniscatic, and pseudo-lemniscatic expressions.

equianharmonic half-periods

$$\omega_1 = \frac{\Gamma^3\left(\frac{1}{3}\right)}{4\pi}\left(\frac{1}{2}(i\sqrt{3}+1)\right) \qquad \omega_2 = \frac{\Gamma^3\left(\frac{1}{3}\right)}{4\pi}$$

lemniscatic half-periods

$$\frac{\mathrm{L}}{2\sqrt{2}} \qquad \frac{\mathrm{L}}{2\sqrt{2}}\,i$$

pseudo-lemniscatic half-periods

$$\frac{\mathrm{L}}{4}(i+1) \qquad \frac{\mathrm{L}}{4}(i-1)$$

The lemniscate plays a central (quadropoly orthogonal) role in this balance because "The lemniscate is the inverse curve of the hyperbola with respect to its center."

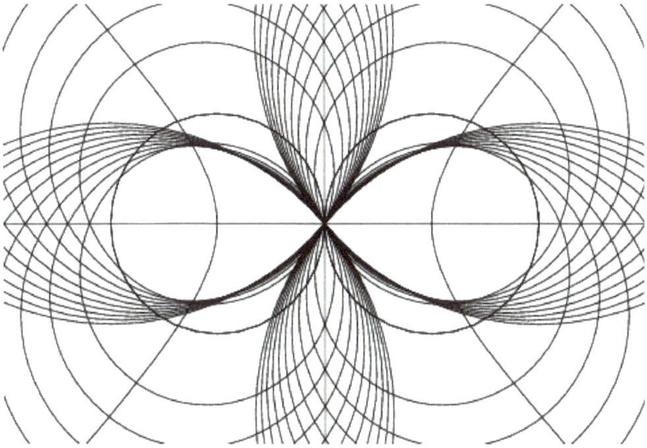

"The lemniscate can also be generated as the envelope of circles centered on a rectangular hyperbola and passing through its center." [3]

Instead of writing balance 2 in terms of the Weierstrass constant, or the Lemniscate constant, for now let's write it in terms of the Gamma function, to highlight how it is maintained under ideally squared split hyperbolic balance.

$$\frac{n}{\sqrt{b\pi}\ (3)^{1/3}} \left(2^{2n}\, e^{\pi} \right)^{-1/8} \left(\Gamma\left(\left(\frac{1}{2} \right)^{2} \right) \right)^{2} e^{\phi_2} = 2(3 \pm 0)^2 \qquad \text{balance 2}$$

Where $\phi_2 = 1.875545967$ ... is the magnitude of the external rotation of this balance, $\Gamma(x)$ = the Gamma function, which encodes hyperbolically balanced partitions, and $2(3 \pm 0)^2 = 18$ = the number of derangements participating in this balance.

[3] Wells, D. *The Penguin Dictionary of Curious and Interesting Geometry.* London: Penguin, pp. 139-140, 1991.

Balance 3 defines the opposite end of hyperbolic vortex, where the partitions of the hyperbolic figure eight knot circularly close $2\pi\, n$ while being squarely balanced on the internal split-division of that knot.

$$2\pi\, n \left(\cos\left(\frac{b}{n}\right)\right)^2 e^{\phi_3} = 2(3-1)^2 = 2^3 \qquad \text{balance 3}$$

Where $\phi_3 = 2.176426838\ldots$ is the magnitude of the external rotation maintaining balance 3, $\cos\left(\frac{b}{n}\right)$ = the internal split-division balance of the hyperbolic figure eight knot, and $2^3 = 8$ = the number of derangements participating in balance 3.

$$\cos\left(\frac{b}{n}\right) = \frac{1}{2}\left(e^{-\left(\frac{b}{n}\right)i} + e^{\left(\frac{b}{n}\right)i}\right)$$

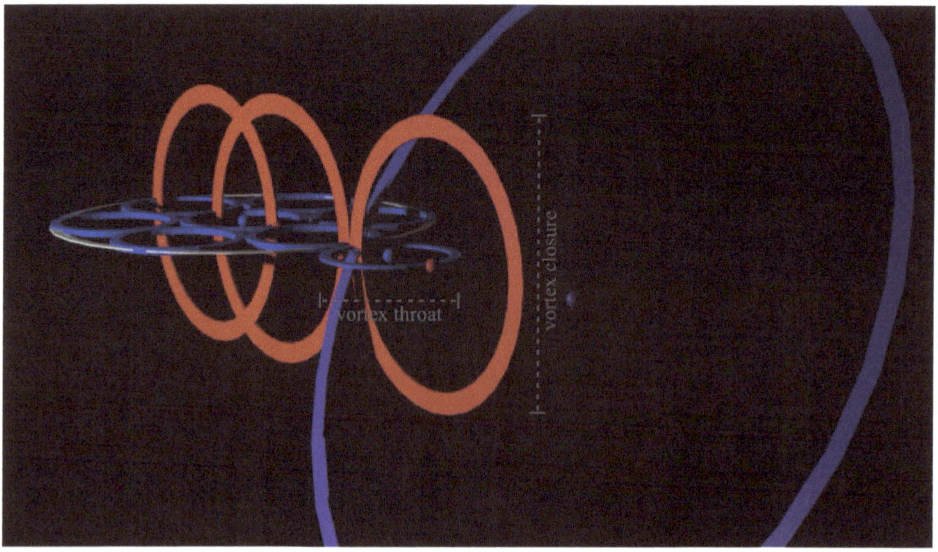

The boundaries of the 2 external balances (balance 2 and 3) form the vortex throat and its terminal scale—the blue and red circles within the external (purple) domain.

The balance point of the external domain (balance 4) prescribes the unitary scale of this decomposition (1)—the scale of stability projected under this action.

The internal boundary of balance 4 is maintained by logarithmically factoring under (you guessed it) ideally squared hyperbolic balance.

$$2\,cosh(log\,b)\left(cosh\left(\frac{n}{2}\right)\right)^2\left(cos\left(\frac{b}{n}\right)\right)^2 e^{\phi_4} = 2(3+1)^2 = 2^n \quad \text{boundary 4}$$

Where $\phi_4 = 1.416786985 \ldots$ is the magnitude of the 4th boundary's rotation, $cosh(x) =$ the hyperbolic cosine function, $log\,x =$ the logarithm function, $cos(x) =$ the cosine function, $n =$ the number of unique rotations partitioning the balance of the hyperbolic figure eight knot, $b = 7$ the break in scale symmetry structurally maintained by that balance, and $2^n = 32 =$ the number of derangements in balance 4.

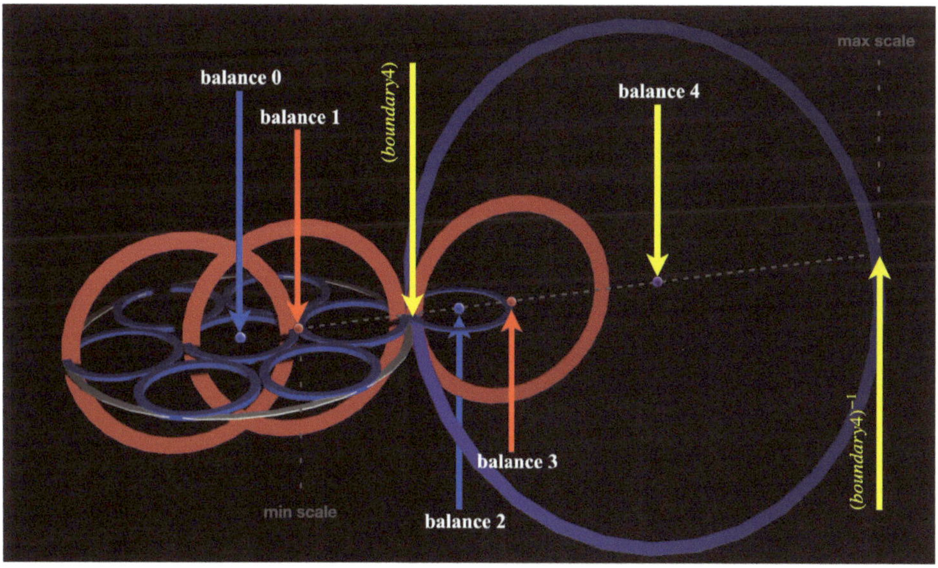

The internal and external boundaries of balance 4 (boundary 4 & (boundary 4)$^{-1}$) are under inverse action.

The external boundary $(\text{Boundary } 4)^{-1}$ inversely expresses the action of boundry 4. This boundary defines the outer-most edge of action in this cyclically defined universe, the other boundary across which no action participating in this balance crosses.

$$(2\,cosh(log\,b))^{-1}\left(cosh\left(\frac{n}{2}\right)\right)^{-2}\left(cos\left(\frac{b}{n}\right)\right)^{-2} e^{-\phi_4} = 2^{-n} \qquad (\text{boundary } 4)^{-1}$$

Where $\phi_4 = 1.416786985 \ldots$ is the magnitude of the 4th boundary's rotation, $cosh(x) =$ the hyperbolic cosine function, $log\,x =$ the logarithm function, $cos(x) =$ the cosine function, $n = 5$ the number of unique rotations partitioning the balance of the hyperbolic figure eight knot, $b = 7$ the break in scale symmetry structurally maintained by the balance of the hyperbolic figure eight knot, and 2^{-n} = the inverse derangements of this balance.

By fully decomposing the partition balance of the hyperbolic figure eight knot we have mapped the most elemental balance possible, defining the simplest possible persistent stage.

the complete partition balance of the hyperbolic figure eight knot

$$(2\,cosh(log\,b))^{-1}\left(cosh\left(\frac{n}{2}\right)\right)^{-2}\left(cos\left(\frac{b}{n}\right)\right)^{-2} e^{-\phi_4} = 2^{-n} \qquad (4^{th})^{-1}$$

$$(\text{external reference}) = 1 \qquad 4^{th}$$

$$2\pi\, n\left(cos\left(\frac{b}{n}\right)\right)^{2} e^{\phi_3} = 2(3-1)^2 = 2^3 \qquad 3^{rd}$$

$$\frac{n}{\sqrt{b\pi}\,(3)^{1/3}}\left(2^{2n}\,e^{\pi}\right)^{-1/8}\left(\Gamma\left(\left(\frac{1}{2}\right)^{2}\right)\right)^{2} e^{\phi_2} = 2(3\pm 0)^2 \qquad 2^{nd}$$

$$2\,cosh(log\,b)\left(cosh\left(\frac{n}{2}\right)\right)^{2}\left(cos\left(\frac{b}{n}\right)\right)^{2} e^{\phi_4} = 2(3+1)^2 = 2^{n} \qquad 4^{th}$$

$$\left(sinh\left(sinh\left(\frac{1}{b}\right)\right)\right)^{-1} e^{\phi_1} = b(n) \qquad 1^{st}$$

$$\pi\left(sinh\left(\left(\frac{1}{2}\right)^{2}\right)\right)^{2} e^{\phi_0} = !n \qquad 0^{th}$$

Where ϕ_k = the external rotation of the kth balance, d_k = the number of derangements participating in the kth balance, $n = 5$, the number of unique rotations partitioning the balance of the hyperbolic figure eight knot, $!n = 44$, the number of derangements available to those 5 rotations, $b = 7$ the break in scale symmetry internally maintained by this balance, $sinh(x)$ = the hyperbolic sine function, $cosh(x)$ = the hyperbolic cosine function $cos(x)$ = the cosine function, and $\Gamma(x)$ = the Gamma function, which encodes hyperbolically balanced partitions.

$\phi_4 = 1.4167869859079\ldots$	$d_4 = 2(3+1)^2 = 2^n = 32$
$\phi_3 = 2.1764268381757\ldots$	$d_3 = 2(3-1)^2 = 2^3 = 8$
$\phi_2 = 1.8755459671396\ldots$	$d_2 = 2(3\pm 0)^2 = 18$
$\phi_1 = 1.6162591817564\ldots$	$d_1 = b(n) = 35$
$\phi_0 = 5.3912583683231\ldots$	$d_0 = !n = 44$

Chapter 5 — the boundary conditions of physical reality

The boundary conditions that maintain the balance of the hyperbolic figure eight knot, define the boundary conditions of physical reality—the measured limits of time, space, charge, mass and temperature known as the Planck constants.

Every persistent external rotation of the hyperbolic figure eight knot, multiplied by a boundary doubling raised to the number of hinges synced to that rotation, defines a bounds of physical measure—a Planck constant.

Planck temperature $T_P = \phi_4(2n)^{+d_4}$

$T_P = 1.4167869859079 \ldots \times 10^{32}$ predicted

$T_P = 1.416784(16) \times 10^{32}\ K$ measured

Planck mass $m_P = \phi_3(2n)^{-d_3}$

$m_P = 2.1764268381757 \ldots \times 10^{-8}$ predicted

$m_P = 2.176434(24) \times 10^{-8}\ kg$ measured

Planck charge $q_P = \phi_2(2n)^{-d_2}$

$q_P = 1.8755459671396 \ldots \times 10^{-18}$ predicted

$q_P = 1.8755459 \times 10^{-18}\ C$ previously defined

Planck length $l_P = \phi_1(2n)^{-d_1}$

$l_P = 1.6162591817564 \ldots \times 10^{-35}$ predicted

$l_P = 1.616255(18) \times 10^{-35}\ m$ measured

Planck time $t_P = \phi_0(2n)^{-d_0}$

$t_P = 5.3912583683231 \ldots \times (10)^{-44}$ predicted

$t_P = 5.391247(60) \times (10)^{-44}\ s$ measured

The hyperbolic figure eight knot defines a bounded arena composed of 5 persistent kinds of measure (time, space, charge, mass, and temperature) whose boundaries are the Planck constants. The Platonic form of persistence, the simplest possible self-balanced manifold, elegantly and uniquely defines the stage of reality.

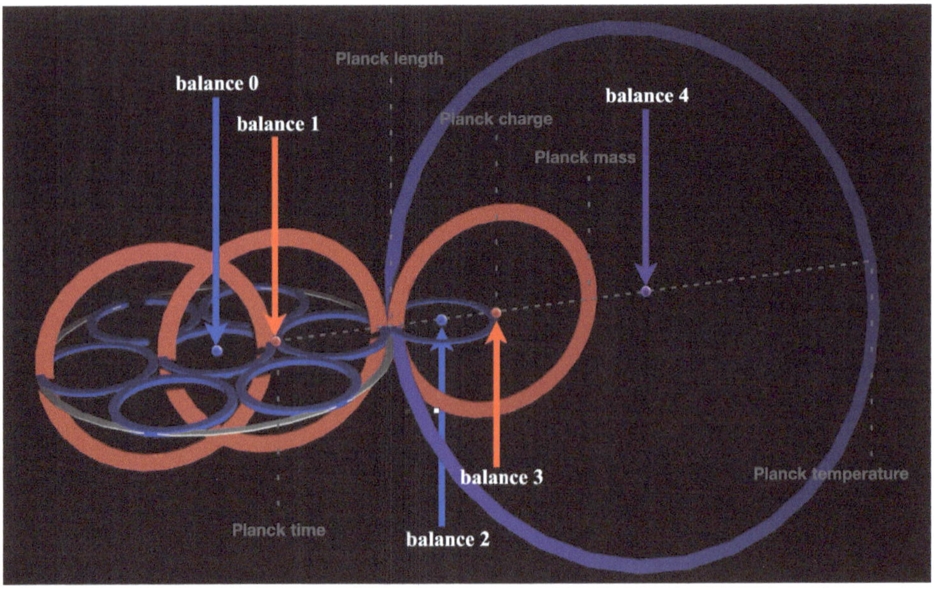

The Planck constants define the action limit (maximum reach/boundary) of each balance (t_P = the maximum reach of balance 0, l_P =the maximum reach of balance 1, etc.)

Chapter 6 — the hyperbolic vortex

The 2 external boundaries of the hyperbolic figure eight knot, the Planck *charge* and Planck *mass* boundaries, connect via the hyperbolic vortex equation.

$$\frac{1}{ж} + ж + \frac{ж^3}{2\pi} = \left(i^i \right)^{-\pi/2} - m_P$$

Where m_P = the Planck mass.

This connection partitions into 4 solutions composed of 6 parts (4 Real and 2 Imaginary).

$$ж_1 = 0.0854245431533310 \ldots$$
$$ж_2 = 3.667567534854999 \ldots$$
$$ж_3 = -1.876496039004165(16) + 4.066152626159671(03)i$$
$$ж_4 = -1.876496039004165(16) - 4.066152626159671(03)i$$

These solutions have ideally cyclic combinatorics.

$ж_1 ж_2 ж_3 ж_4 = 2\pi$ product

$ж_1 + ж_2 + ж_3 + ж_4 = 0$ sum

$ж_1{}^2 + ж_2{}^2 + ж_3{}^2 + ж_4{}^2 = -4\pi$ square sum

$Re(ж_3) = Re(ж_4)$ twin

$Im(ж_3) = -Im(ж_4)$ reflection

$\left(Re(ж_3)\right)^2 + \left(Im(ж_3)\right)^2 = \left(Re(ж_4)\right)^2 + \left(Im(ж_4)\right)^2 = ж_3 ж_4$ pair

$\left(Re(ж_3)\right)^2 + \left(Im(ж_4)\right)^2 = \left(Re(ж_4)\right)^2 + \left(Im(ж_3)\right)^2 = ж_3 ж_4$ pair

$$\sum_{k=1}^{4} \text{ж}_k = 0 \qquad \prod_{k=1}^{4} \text{ж}_k = 2\pi$$

$$\sum_{k=1}^{4} {\text{ж}_k}^2 = -4\pi \qquad \prod_{k=1}^{4} {\text{ж}_k}^2 = 4\pi^2$$

$$-\pi \sum_{k=1}^{4} {\text{ж}_k}^2 = \prod_{k=1}^{4} {\text{ж}_k}^2$$ hyperbolic vortex partition relation

Since these solutions define the connection between the Planck charge and Planck mass boundaries, they define the natural partitions of *charge* and *mass*.

Chapter 7 — charge partitions

When charge is externally present (charge = 0 at the Planck temperature boundary and the inverse Planck temperature boundary) its magnitude is equal to the charge boundary (the Planck charge q_P) multiplied by the first solution of the hyperbolic vortex equation ($ж_1$).

$$e = ж_1 q_P$$

Internally this charge partitions exactly as the volume of the hyperbolic figure eight knot partitions (in the imaginary direction), into a negative one-thirds and positive two-thirds action (dilogarithmic flip).

$$V_{fe} = i\left[-Li_2\left((-1)^{\frac{1}{3}}\right) + Li_2\left((-1)^{\frac{2}{3}}\right)\right]$$

Where Li_2 is the hyperbolic dilogarithm.

This arranges charge into the following fundamental partitions.

charge partitions

$(Z)\ (H)\ (W)$	$\leftrightarrow$	$(\pm1)\ (0)\ (\pm1)$
$(P)\ (N)\ (e)$	$\leftrightarrow$	$(+1)\ (0)\ (-1)$
$(\nu_\tau)\ (\nu_\mu)\ (\nu_e)$	$\leftrightarrow$	$(0)\ (0)\ (0)$

external

- -

internal

$(d)\ (e)\ (u)$	$\leftrightarrow$	$\left(-\frac{1}{3}\right)(-1)\left(+\frac{2}{3}\right)$
$(s)\ (\mu)\ (c)$	$\leftrightarrow$	$\left(-\frac{1}{3}\right)(-1)\left(+\frac{2}{3}\right)$
$(b)\ (\tau)\ (t)$	$\leftrightarrow$	$\left(-\frac{1}{3}\right)(-1)\left(+\frac{2}{3}\right)$

Where $e = ж_1 q_P$ is set as the scale of charge unity = 1, and $H, Z, W, N, P, e, \nu_\tau, \nu_\mu, \nu_e, d, u, s, c, \mu, b, t,$ and τ are the charge values of the Higgs boson, Z boson, W boson, neutron, proton, electron, down quark, up quark, charm quark, strange quark, muon, truth (top) quark, tau quark, beauty (bottom) quark, tau neutrino, mu neutrino, and the electron neutrino.

Chapter 8 — mass partitions

The mass boundary binds 2 figure eight knot volumes in 4 orthogonal ways, and trivially terminates inverse-hyperbolically over the derangement structure of the hyperbolic figure eight knot. This primitive tessellation of the mass boundary defines the mass of the electron (m_e).

$$m_e = 2V_{fe}\, m_P{}^4 \left(1 + \left(sinh\left(sinh\left(\frac{!\,n}{b}\right)\right)^{-1}\right)^{-1} \left(\frac{l_P\, m_P}{q_P{}^2}\right)\right)$$

$m_e = 9.10938370161994 \ldots \times 10^{-31}\, kg$ predicted

$m_e = 9.1093837015(28) \times 10^{-31}\, kg$ measured

Where V_{fe} = the complement volume of the hyperbolic figure eight knot, m_P = the Planck mass boundary, $sinh(x)$ = the hyperbolic sine function, n = 5 the number of unique rotations the hyperbolic figure eight knot partitions into, $!\,n$ = 44 the number of derangements available to those partitions, b = 7 the break in scale symmetry internally maintained by the balance of the hyperbolic figure eight knot, and $\left(\frac{l_P\, m_P}{q_P{}^2}\right)$ = the termination boundary of the external split metric.

It is instructive to note that the reflection, or external action, of every inverted cyclic balance maintained within the hyperbolic figure eight knot terminates on this boundary $\left(\frac{l_P\, m_P}{q_P{}^2}\right)$, because this is where the sum of its 5 external rotations collectively partition into the n-hypersphere of maximal volume.

$$\sum_{k=0}^{4} \phi_k = \left(\frac{2}{3}\right)^2 \frac{4\pi}{Re(\omega_1)^3}\left(1 + \left(\frac{3}{2}\right)^2 \left(\frac{V_h^*}{n}\right)\left(\frac{l_P\, m_P}{q_P{}^2}\right)\right) \quad \text{total rotation}$$

Where V_h^* = the dimension at which an n-hypersphere has maximal volume, A_h^* = the dimension at which an n-hypersphere has maximal surface area,

$$V_h^* - A_h^* = \left(\frac{2}{3}\right)(b - 2n) \quad \text{hypersphere max volume to surface area}$$

Any squarely balanced (periodic) hyperbolic space has a total of 17 possible tessellations, or tilings. (There are exactly 17 plane symmetry groups.)

To find all of the possible tessellations of the *mass* boundary maintained by the balance of the hyperbolic figure eight knot, we orthogonally extend its trivial tessellation (m_e) under ideal balance of (splitting)^2 and joining—guaranteeing that:

1. the Planck charge and Planck mass boundaries connect via the hyperbolic vortex equation, and
2. the balance of that connection squarely hyperbolically bifurcates (splits) in 2 orthogonal but otherwise indistinguishable ways, while
3. dilogarithmically persisting under a unitary hyperbolic-circular balance (being braided under the $+\left(\frac{2}{3}\right)$ and $-\left(\frac{1}{3}\right)$ internal split action of the hyperbolic figure eight knot).

The tessellations that orthogonally capture this balance of *square splitting* and *joining* define the proton mass (m_+) and the neutron mass (m_N).

$$\left(\frac{m_e}{m_+}\right)\left(\frac{ж_2}{ж_1}\right)^2 = \left(\frac{2}{3}\right)^2 (\alpha_F - 1)^2 \left(1 - \left(\frac{1}{3}\right) e^{3\gamma}\left(\frac{l_P\, m_P}{{q_P}^2}\right)\right) \quad \text{splitting map}$$

$$\left(\frac{m_N - m_+}{m_e}\right) = \left(\frac{1}{3}\right)(\mu + 3 + \pi)\left(1 + \left(\frac{2}{3}\right) e^{3\gamma}\left(\frac{l_P\, m_P}{{q_P}^2}\right)\right) \quad \text{joining map}$$

$m_+ = 1.67262192371195 \ldots \times 10^{-27}\, kg$ predicted
$m_+ = 1.67262192369(51) \times 10^{-27}\, kg$ measured

$m_N = 1.67492749802284 \ldots \times 10^{-27}\, kg$ predicted
$m_N = 1.67492749804(95) \times 10^{-27}\, kg$ measured

Where m_+ = the proton mass, m_N = the neutron mass, α_F = the alpha Feigenbaum constant, the 2nd period-doubling bifurcation constant of the logistic map, μ = the nontrivial zero of the logarithmic integral γ = the Euler-Mascheroni constant, the limiting difference between the harmonic series and the logarithm function, and $ж_1$ & $ж_2$ = the 1st and 2nd solutions to the hyperbolic vortex equation.

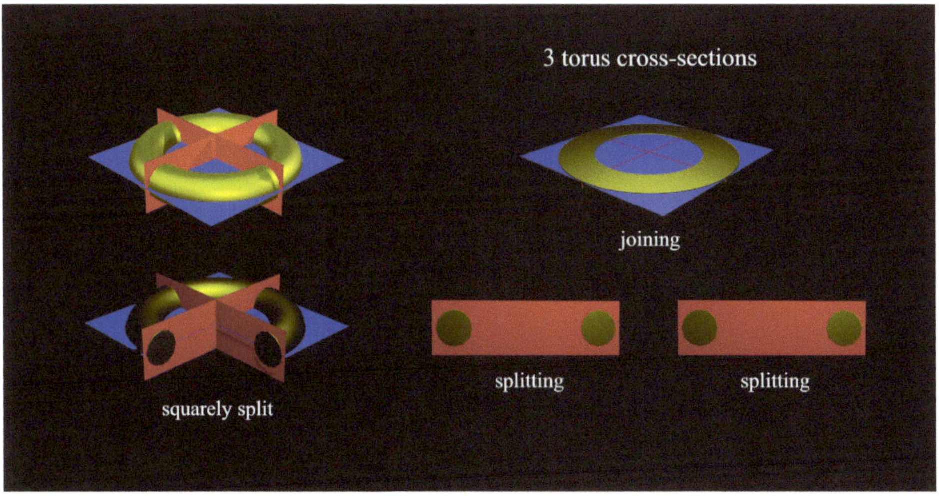

The hyperbolic figure eight knot externally fibers under 2 orthogonal circular actions which trace the boundary of a once-punctured torus. That boundary defines the orthogonal intersection of 2 squarely arranged circle-splittings and 1 joining of circles.

Mirroring the internal division structure of the hyperbolic figure eight knot's complement volume, which is composed of 6 circles surrounding a central circle that successively undergoes 4 orthogonally arranged bifurcations,

$$= \left(\frac{m_H - m_Z}{m_W}\right)6$$

$$= \left(\frac{m_N - m_+}{m_e}\right)$$

$$= \left(\frac{m_d - m_u}{m_e}\right)\left(\frac{1}{2}\right)$$

$$= \left(\frac{m_c - 2m_s}{m_\mu}\right)\left(\frac{1}{2}\right)^2$$

$$= \left(\frac{m_t - m_\tau}{m_b}\right)\left(\left(\frac{1}{2}\right)^2\right)^2$$

$$= \left(\frac{m_{\nu_\tau} - m_{\nu_\mu}}{m_{\nu_e}}\right)\left(\left(\left(\frac{1}{2}\right)^2\right)^2\right)^2$$

the joining map of the 3 trivial tessellations (m_e, m_+, m_N) extends into a balance of externally phased and internally folded tessellations. And the orthogonally connected splitting map extends into a balance of twists, splits and flips. These extensions simultaneously capture all 17 mass tilings.

full joining map

$$\left(\frac{m_H - m_Z}{m_W}\right) = \left(\frac{1}{3}\right)^2 (\mu + 3 + \pi)(2^{-1})\left(1 + \left(\frac{ж_2{}^2}{ж_1{}^3}\right) G_{Ga}{}^{-3}\left(\frac{n}{b}\right)\left(\frac{l_P\, m_P}{q_P{}^2}\right)\right)$$

$$\left(\frac{m_N - m_+}{m_e}\right) = \left(\frac{1}{3}\right) (\mu + 3 + \pi)(2^0)\left(1 + \left(\frac{2}{3}\right) e^{3\gamma}\left(\frac{l_P\, m_P}{q_P{}^2}\right)\right)$$

$$\left(\frac{m_d - m_u}{m_e}\right) = \left(\frac{1}{3}\right) (\mu + 3 + \pi)(2^{+1})\left(1 + \left(\frac{ж_2{}^2}{ж_1{}^3}\right) D_{Do}{}^3 \sqrt{\frac{2}{b}}\left(\frac{l_P\, m_P}{q_P{}^2}\right)\right)$$

external phases

- -

internal folds

$$\left(\frac{m_c - 2m_s}{m_\mu}\right) = \left(\frac{1}{3}\right) (\mu + 3 + \pi)(2^2)\left(1 + 0\left(\frac{l_P\, m_P}{q_P{}^2}\right)\right)$$

$$\left(\frac{m_t - m_\tau}{m_b}\right) = \left(\frac{1}{3}\right) (\mu + 3 + \pi)\left(2^{2^2}\right)\left(1 + b\frac{\sqrt{Re(\omega_1)}}{ж_1{}^4}\left(\frac{l_P\, m_P}{q_P{}^2}\right)\right)$$

$$\left(\frac{m_{\nu_\tau} - m_{\nu_\mu}}{m_{\nu_e}}\right) = \left(\frac{1}{3}\right) (\mu + 3 + \pi)\left(2^{2^{2^2}}\right)\left(1 + 0\left(\frac{l_P\, m_P}{q_P{}^2}\right)\right)$$

Where m_H, m_Z, m_W, m_N, m_P, m_e, m_d, m_u, m_c, m_s, m_μ, m_t, m_τ, m_b, m_{ν_τ}, m_{ν_μ}, and m_{ν_e} are the masses of the Higgs boson, Z boson, W boson, neutron, proton, electron, down quark, up quark, charm quark, strange quark, muon, truth (top) quark, tau quark, beauty (bottom) quark, tau neutrino, mu neutrino, and the electron neutrino; μ is the Ramanujan-Soldner constant (the nontrivial zero of the logarithmic integral), γ is the Euler-Mascheroni constant (which defines the limiting difference between the natural logarithm and the harmonic series), $ж_1$ and $ж_2$ are the first and second Real solutions to the vortex equation, G_{Ga} is Gauss's constant, D_{Do} is the unique real root of the cosine function (a universal attracting fixed point called the Dottie number), and ω_1 is the omega_1 constant.

full splitting map

$$\frac{m_b + m_c + m_t}{\left(\sqrt{m_b} + \sqrt{m_c} + \sqrt{m_t}\,\right)^2} = \left(\frac{2}{3}\right)^{3} (\alpha_F - 1)^{2} \left(1 + \left(\frac{\text{ж}_2{}^3}{\text{ж}_1{}^2}\right) \frac{V_{fe}{}^4}{2^{n}} \left(\frac{l_P\, m_P}{q_P{}^2}\right)\right)$$

$$\left(\frac{m_e}{m_+}\right)\left(\frac{\text{ж}_2}{\text{ж}_1}\right)^2 = \left(\frac{2}{3}\right)^{2} (\alpha_F - 1)^{2} \left(1 - \left(\frac{1}{3}\right) e^{3\gamma} \left(\frac{l_P\, m_P}{q_P{}^2}\right)\right)$$

$$\frac{m_e + m_\mu + m_\tau}{\left(\sqrt{m_e} + \sqrt{m_\mu} + \sqrt{m_\tau}\,\right)^2} = \left(\frac{2}{3}\right)^{1} (\alpha_F - 1)^{0} \left(1 - (3\,\text{ж}_2{}^2)\, P_{up} \left(\frac{l_P\, m_P}{q_P{}^2}\right)\right)$$

twist-split balance

- -

flip-split balance

$$\frac{m_H + m_Z + m_W}{\left(\sqrt{m_H} + \sqrt{m_Z} + \sqrt{m_W}\,\right)^2} = \left(\frac{2}{3}\right)^{-\frac{1}{2}} (\alpha_F - 1)^{-e^{2\gamma}} \left(1 + \left(\frac{\text{ж}_2}{\text{ж}_1}\right)^2 \left(2nb\,(V_{fe})^{\frac{1}{3}}\right)^{-\frac{1}{2}} \left(\frac{l_P\, m_P}{q_P{}^2}\right)\right)$$

$$\frac{m_u + m_s + m_d}{\left(\sqrt{m_u} + \sqrt{m_s} + \sqrt{m_d}\,\right)^2} = \left(\frac{2}{3}\right)^{0} \gamma\, (\alpha_F - 1)^{0} \left(1 - 0 \left(\frac{l_P\, m_P}{q_P{}^2}\right)\right)$$

$$\frac{m_{\nu_e} + m_{\nu_\mu} + m_{\nu_\tau}}{\left(\sqrt{m_{\nu_e}} + \sqrt{m_{\nu_\mu}} + \sqrt{m_{\nu_\tau}}\,\right)^2} = \left(\frac{2}{3}\right)^{-\frac{1}{2}} (\alpha_F - 1)^{+e^{\pi i}} \left(1 - \left(\frac{\text{ж}_2}{\text{ж}_1}\right)^2 \left(2nb\,(V_{fe})^{\frac{2}{3}}\right)^{-\frac{1}{2}} \left(\frac{l_P\, m_P}{q_P{}^2}\right)\right)$$

Where m_H, m_Z, m_W, m_N, m_P, m_e, m_d, m_u, m_c, m_s, m_μ, m_t, m_τ, m_b, m_{ν_τ}, m_{ν_μ}, and m_{ν_e} are respectively the masses of the: Higgs boson, Z boson, W boson, neutron, proton, electron, down quark, up quark, charm quark, strange quark, muon, truth quark, tau quark, beauty quark, tau neutrino, mu neutrino, and the electron neutrino, α_F is the alpha Feigenbaum constant, γ is the Euler-Mascheroni constant (which denotes the limiting difference between the natural logarithm and the harmonic series), ж_1, ж_2 are the first and second Real solutions to the vortex equation, V_{fe} is the figure eight knot volume complement, and P_{up} is the universal parabolic constant.

Chapter 9 — individual tessellations

Every unique tessellation of this mass boundary represents a fundamental geometric tiling from which the hyperbolic figure eight knot's external *mass boundary* can be periodically constructed. These tilings are either geometrically combinatorial (composed of an inverted balance stabilized between 2 actions, the second of which is always on the termination boundary), or primitive (referencing a single geometric action).

Let's examine the geometry of each tiling.

charm quark mass

$$\frac{m_e}{m_c}\left(\frac{ж_2}{ж_1}\right)^2 = D_{DO}$$

Where D_{DO} = the Dottie number, the unique real fixed point of the cosine function—a universal attracting point.

$m_c = 2.27188026398178 \ldots \times 10^{-27}\ kg$ predicted
$m_c = 2.272(63) \times 10^{-27}\ kg$ measured

down quark mass

$$\frac{m_e}{m_d} = \frac{\mu}{ж_2{}^2}$$

Where μ = the nontrivial zero of the logarithmic integral

$m_d = 8.44242715614137 \ldots \times 10^{-30}\ kg$ predicted
$m_d = 8.3(07) \times 10^{-30}\ kg$ measured

up quark mass

$$\frac{m_e}{m_u} = \frac{2}{\omega_2{}^n}$$

Where ω_2 = the omega_2 constant

$$\omega_2 = \frac{\Gamma^3\left(\frac{1}{3}\right)}{4\pi}$$

$m_u = 3.81810683898335 \ldots \times 10^{-30}\ kg$ predicted
$m_u = 3.92(89) \times 10^{-30}\ kg$ measured

strange quark mass

$$\frac{m_e}{m_s}\left(\frac{!n}{\sqrt{n}}\right) = \frac{\sqrt{K_{\_1}}}{ж_2{}^2}$$

Where $K_{\_1}$ = the Khinchin harmonic mean

$m_s = 1.82501207639326 \ldots \times 10^{-28}\ kg$ predicted
$m_s = 1.69(16) \times 10^{-28}\ kg$ measured

bottom quark mass

$$\frac{m_e}{m_b}\left(\frac{ж_2}{ж_1}\right)^2 = \frac{{L_1}^3}{2n}$$

Where L_1 = the 1st Lemniscate constant

m_b = 7.45149186313980 ... × 10^{-27} kg predicted
m_b = 7.45(07) × 10^{-27} kg measured

top quark mass

$$\frac{m_e}{m_t}\left(\frac{{ж_2}^3}{{ж_1}^2}\right) = \frac{L_2}{6n}$$

Where L_2 = the 2nd Lemniscate constant

m_t = 3.08390948667753 ... × 10^{-25} kg predicted
m_t = 3.084(07) × 10^{-25} kg measured

Higgs mass

$$\frac{m_e}{m_H}\left(\frac{ж_2}{ж_1}\right)^2 = \frac{b}{2\,{j_{0,1}}^b}$$

m_H = 2.23150010999262 ... × 10^{-25} kg predicted
m_H = 2.2315(28) × 10^{-25} kg measured

Z boson mass

$$\frac{m_e}{m_Z}\left(\frac{{ж_2}^3}{{ж_1}^2}\right) = \left(\frac{2}{3}\right)^3 {L_{LL}}^n$$

Where L_{LL} = the Laplace limit, $\frac{1}{L_{LL}} = \sinh(C_{CFP})$, and C_{CFP} = the real fixed point of the hyperbolic cotangent

m_Z = 1.62556627846185 ... × 10^{-25} kg predicted
m_Z = 1.625566(38) × 10^{-25} kg measured

W boson mass

$$\frac{m_e}{m_W}\left(\frac{ж_2{}^2}{ж_1{}^3}\right) = \left(Im(\rho_1)\right)^{-3/4}$$

Where ρ_1 = the first nontrivial root of the zeta function

$m_W = 1.43263881046217 \ldots \times 10^{-25}\ kg$ predicted
$m_W = 1.43288(21) \times 10^{-25}\ kg$ measured

tau mass

$$\frac{m_e}{m_\tau}\left(\frac{ж_2}{ж_1}\right)^2 = \left(\frac{3}{V_{fe}{}^2}\right)^2 \left(1 - ж_2{}^2 \operatorname{sech}\left(\tan\left(\frac{1}{2}\right)\right)\left(\frac{l_P\, m_P}{q_P{}^2}\right)\right)$$

$m_\tau = 3.16754001786349 \ldots \times 10^{-27}\ kg$ predicted
$m_\tau = 3.16754(21) \times 10^{-27}\ kg$ measured

muon mass

$$\frac{m_e}{m_\mu} = ж_1{}^2\, L_{LL}\left(1 + \frac{b\, ж_2}{V_{fe}{}^2}\left(\frac{!n}{\sqrt{n}}\right)\left(\frac{l_P\, m_P}{q_P{}^2}\right)\right)$$

Where L_{LL} = the Laplace limit, $\frac{1}{L_{LL}} = \sinh(C_{CFP})$, and C_{CFP} = the real fixed point of the hyperbolic cotangent

$m_\mu = 1.88353162775 4459 \ldots \times 10^{-28}\ kg$ predicted
$m_\mu = 1.883531627(42) \times 10^{-28}\ kg$ measured

tau neutrino mass

$$\frac{\phi_1(2n)^{-bn}}{m_{\nu_\tau}}\left(\frac{ж_1}{ж_2}\right)^2 = 2^{1/n}\left(1 - (\sqrt{2} - 1)\left(\frac{ж_2}{ж_1}\right)^2\left(\frac{l_P\, m_P}{q_P{}^2}\right)\right)$$

$m_{\nu_\tau} = 7.63391385156818 \ldots \times 10^{-39}\ kg$ predicted
$m_{\nu_\tau} = ???\ kg$ no current measurement

muon neutrino mass

$$m_{\nu_\mu} = 3\ \phi_0 (2n)^{-!n}\ kg$$

$m_{\nu_\mu} = 1.61737751049693 \ldots \times 10^{-43}\ kg$ predicted

$m_{\nu_\mu} = ???\ kg$ no current measurement

electron neutrino mass

$$m_{\nu_e} = 2\ \phi_0 (2n)^{-!n}\ kg$$

$m_{\nu_e} = 1.07825167366462 \ldots \times 10^{-43}\ kg$ predicted

$m_{\nu_e} = ???\ kg$ no current measurement

This updates our list of hyperbolic figure eight knot partition parameters to the following.

the partition parameters of the hyperbolic figure eight knot

$n = 5$ number of unique rotations
$b = 7$ break in scale symmetry

$\phi_0 = 5.39125836832313 \ldots + 2\pi i(k) \quad k \in \mathbb{Z}$ 0th external rotation
$\phi_1 = 1.61625918175645 \ldots + 2\pi i(k)$ 1st external rotation
$\phi_2 = 1.87554596713962 \ldots + 2\pi i(k)$ 2nd external rotation
$\phi_3 = 2.17642683817579 \ldots + 2\pi i(k)$ 3rd external rotation
$\phi_4 = 1.41678698590795 \ldots + 2\pi i(k)$ 4th external rotation

$t_P = 5.39125836832313 \ldots \times 10^{-44}\ s$ Planck time
$l_P = 1.61625918175645 \ldots \times 10^{-35}\ m$ Planck length
$q_P = 1.87554596713962 \ldots \times 10^{-18}\ C$ Planck charge
$m_P = 2.17642683817579 \ldots \times 10^{-8}\ kg$ Planck mass
$T_P = 1.41678698590795 \ldots \times 10^{32}\ K$ Planck temperature

$V_{fe} = 2.02988321281930 \ldots$ figure eight knot complement volume
$e = 2.71828182845904$ Euler's number
$\pi = 3.14159265358979 \ldots$ Archimedes' constant
$ж_1 = 0.0854245431533304 \ldots$ 1st hyperbolic vortex constant
$ж_2 = 3.66756753485499 \ldots$ 2nd hyperbolic vortex constant
$ж_3 = -1.87649603900417 \ldots + 4.06615262615972 \ldots i$ 3rd
$ж_4 = -1.87649603900417 \ldots - 4.06615262615972 \ldots i$ 4th
$\alpha_F = 2.50290787509589 \ldots$ alpha Fiegenbaum constant
$\delta_F = 4.66920160910299 \ldots$ delta Fiegenbaum constant
$\gamma = 0.577215664901532 \ldots$ Euler-Mascheroni constant
$\mu = 1.45136923488338 \ldots$ nontrivial zero of the logarithmic integral
$G_{Ga} = 0.834626841674073 \ldots$ Gauss's constant
$L = 2.622057554292119 \ldots$ Lemniscate constant
$L_1 = 1.31102877714605 \ldots$ 1st Lemniscate constant
$L_2 = 0.599070117367796 \ldots$ 2nd Lemniscate constant
$D_{Do} = 0.739085133215160 \ldots$ Dottie number
$\omega_1 = 0.764977018528596 \ldots + 1.32497062714087 \ldots i$ omega_1
$\omega_2 = 1.529954037057192 \ldots$ omega_2 constant
$P_{up} = 2.29558714939263 \ldots$ universal parabolic constant
$K_{\_1} = 1.745405662407346 \ldots$ Khinchin harmonic mean
$j_{0,1} = 2.40482555769577 \ldots$ 1st root of the Bessel function
$L_{LL} = 0.662743419349181 \ldots$ Laplace limit
$C_{CFP} = 1.19967864025773 \ldots$ real fixed point of the hyperbolic cotangent
$\rho_1 = 0.5 + 14.1314251417346 \ldots i$ 1st nontrivial zero of the zeta function

$m_e = 9.10938370161994 \ldots \times 10^{-31}\ kg$ electron mass
$m_+ = 1.67262192371195 \ldots \times 10^{-27}\ kg$ proton mass
$m_N = 1.67492749802284 \ldots \times 10^{-27}\ kg$ neutron mass
$m_c = 2.27188026398178 \ldots \times 10^{-27}\ kg$ charm quark mass
$m_d = 8.44242715614137 \ldots \times 10^{-30}\ kg$ down quark mass
$m_u = 3.81810683898335 \ldots \times 10^{-30}\ kg$ up quark mass
$m_s = 1.82501207639326 \ldots \times 10^{-28}\ kg$ strange quark mass
$m_b = 7.45149186313980 \ldots \times 10^{-27}\ kg$ beauty (bottom) quark mass
$m_t = 3.08390948667753 \ldots \times 10^{-25}\ kg$ truth (top) quark mass
$m_H = 2.23150010999262 \ldots \times 10^{-25}\ kg$ Higgs boson mass
$m_Z = 1.62556627846185 \ldots \times 10^{-25}\ kg$ Z boson mass
$m_W = 1.43263881046217 \ldots \times 10^{-25}\ kg$ W boson mass
$m_\tau = 3.16754001786349 \ldots \times 10^{-27}\ kg$ tau mass
$m_\mu = 1.88353162775445 \ldots \times 10^{-28}\ kg$ muon mass
$m_{\nu_\tau} = 7.63391385156818 \ldots \times 10^{-39}\ kg$ tau neutrino mass
$m_{\nu_\mu} = 1.61737751049693 \ldots \times 10^{-43}\ kg$ muon neutrino mass
$m_{\nu_e} = 1.07825167366462 \ldots \times 10^{-43}\ kg$ electron neutrino mass

Where the black digits represent previously known values (either measured or geometrically known), and the green digits represent predictions.

Chapter 10 — the constants of Nature

So far we have found that the balance of the hyperbolic figure eight knot is responsible for the persistent arena of reality, minimally decomposing into 5 unique actions (time, space, charge, mass and temperature) with sharply defined boundaries (the Planck constants). The 2 external expressions of that balance (its *charge* and *mass* boundaries) connect via the hyperbolic vortex equation. And the 17 possible tessellations of that connection define the charge and mass values of all the fundamental particles of matter.

In this Chapter, we identify the remaining actions of this partition balance, revealing the geometry that defines the balance of each unique boundary intersection within the hyperbolic figure eight knot. We discover that every unique partition balance defines a constant of Nature. Therefore, every constant of Nature is uniquely defined in terms of the hyperbolic figure eight knot's partition constants, and uniquely defines one of the balances of those elementary partitions.

balance template

constant of Nature:
Compton wavelength

primary boundary　　terminal boundary

$$\lambda_c = 2\pi \left(\frac{l_P m_P}{m_e} \right) \left(1 + \alpha_F \left(\frac{b}{2} \right) \left(\frac{l_P m_P}{q_P^2} \right) \right)$$

primary action　　terminal action

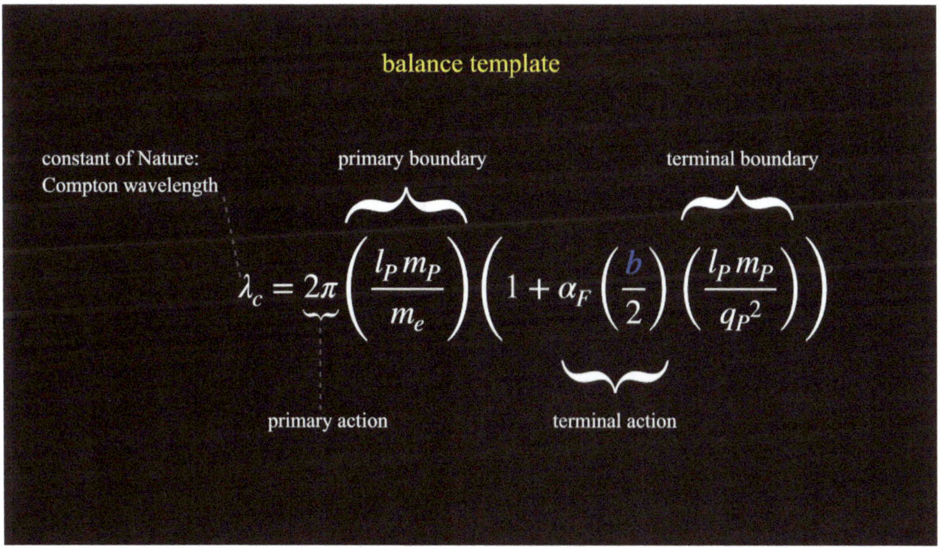

Every inverted balance joins a primary geometric action with a geometric action on the fixed (always the same) terminal boundary.

fine-structure constant

$$\alpha = ж_1{}^2$$

$\alpha = 7.2973525729552\ldots \times 10^{-3}$ predicted
$\alpha = 7.2973525698(24) \times 10^{-3}$ measured

elementary charge

$$e = ж_1 q_P$$

$e = 1.6021765740597\ldots \times 10^{-19}\ C$ predicted
$e = 1.602176565(35) \times 10^{-19}\ C$ measured

Compton wavelength

$$\lambda_C = 2\pi\left(\frac{l_P\, m_P}{m_e}\right)\left(1 + \alpha_F\left(\frac{b}{2}\right)\left(\frac{l_P\, m_P}{q_P{}^2}\right)\right)$$

$\lambda_C = 2.42631023893 94\ldots \times 10^{-12}\ m$ predicted
$\lambda_C = 2.4263102389(16) \times 10^{-12}\ m$ measured

Josephson constant

$$K_J = \frac{ж_1}{\pi}\left(\frac{t_P\, q_P}{l_P{}^2\, m_P}\right)\left(1 - \left(\frac{e^{2\gamma}}{Im(\omega_1)^3}\right)\left(\frac{b}{2}\right)\left(\frac{l_P\, m_P}{q_P{}^2}\right)\right)$$

$K_J = 4.835978484 00467\ldots \times 10^{14}\ \ sC/m^2kg$ predicted
$K_J = 4.835978484 \times 10^{14}\ \ sC/m^2kg$ previously defined

Planck's constant

$$\hbar = \left(\frac{l_P{}^2\, m_P}{t_P}\right)\left(1 + \sqrt{\frac{b\pi}{\zeta(3)}}\left(\frac{l_P\, m_P}{q_P{}^2}\right)\right)$$

Where $\zeta(3)$ = Apéry's constant

$\hbar = 1.05457172593010\ldots \times 10^{-34}\ \ m^2kg/s$ predicted
$\hbar = 1.054571726(47) \times 10^{-34}\ \ m^2kg/s$ measured

electric constant

$$\varepsilon_0 = \frac{1}{4\pi}\left(\frac{t_P{}^2\, q_P{}^2}{l_P{}^3\, m_P}\right)\left(1 - \left(\frac{n^2}{2^n}\right)\left(\frac{l_P\, m_P}{q_P{}^2}\right)\right)$$

$\varepsilon_0 = 8.85418781308692 \ldots \times 10^{-12}\ s^2C^2/m^3kg$ predicted
$\varepsilon_0 = 8.8541878128(13) \times 10^{-12}\ s^2C^2/m^3kg$ measured

Coulomb's constant

$$\kappa = \left(\frac{l_P{}^3\, m_P}{t_P{}^2\, q_P{}^2}\right)\left(1 + \left(\frac{n^2}{2^n}\right)\left(\frac{l_P\, m_P}{q_P{}^2}\right)\right)$$

$\kappa = 8.98755179196986 \ldots \times 10^9\ m^3kg/s^2C^2$ exact
$\kappa = 8.9875517923(14) \times 10^9\ m^3kg/s^2C^2$ measured

quantized Hall conductance

$$H_C = \frac{ж_1{}^2}{2\pi}\left(\frac{t_P\, q_P{}^2}{l_P{}^2 m_P}\right)\left(1 - ж_2\left(\frac{l_P\, m_P}{q_P{}^2}\right)\right)$$

$H_C = 3.87404614816855 \times 10^{-5}\ C^2/m^2kg$ predicted
$H_C = 3.87404614(17) \times 10^{-5}\ C^2/m^2kg$ measured

Bohr magneton

$$\mu_B = \frac{ж_1}{2}\left(\frac{l_P{}^2\, q_P\, m_P}{t_P\, m_e}\right)\left(1 + \frac{2\pi^2}{ж_2}\left(\frac{l_P\, m_P}{q_P{}^2}\right)\right)$$

Note: $2\pi^2/ж_2 = \pi\, ж_1ж_3ж_4$

$\mu_B = 9.27400994938886 \ldots \times 10^{-24}\ m^2C/s$ predicted
$\mu_B = 9.274009994(57) \times 10^{-24}\ m^2C/s$ measured

magnetic constant

$$\mu_0 = 4\pi \left(\frac{l_P\, m_P}{q_P^{\,2}}\right)\left(1 + ж_2\, e^{\gamma} \sqrt{\frac{3}{2}\left(\frac{l_P\, m_P}{q_P^{\,2}}\right)}\right)$$

$\mu_0 = 1.25663706143747\ldots \times 10^{-6}\ mkg/C^2$ predicted
$\mu_0 = 1.256637061\ldots \times 10^{-6}\ mkg/C^2$ previously defined

1st radiation constant

$$c_1 = 4\pi^2 \left(\frac{l_P^{\,4}\, m_P}{t_P^{\,3}}\right)\left(1 - ж_2^{\,2}(\,2\gamma - 1\,)\left(\frac{l_P\, m_P}{q_P^{\,2}}\right)\right)$$

$c_1 = 3.74177185217629\ldots \times 10^{-16}\ m^4kg/s^3$ predicted
$c_1 = 3.741771852\ldots \times 10^{-16}\ m^4kg/s^3$ previously defined

electron Thomson cross section

$$\sigma_e = \left(\frac{2}{3}\right) 4\pi \left(ж_1^{\,2}\,\frac{l_P\, m_P}{m_e}\right)^2 \left(1 + ж_2^{\,2}(\,L - 1\,)\left(\frac{l_P\, m_P}{q_P^{\,2}}\right)\right)$$

Where L = the Lemniscate constant

$\sigma_e = 6.65246159951664\ldots \times 10^{-29}\, m^2$ predicted
$\sigma_e = 6.6524616(18) \times 10^{-29}\, m^2$ measured

conductance quantum

$$G_0 = \frac{ж_1^{\,2}}{\pi}\left(\frac{t_P\, q_P^{\,2}}{l_P^{\,2} m_P}\right)\left(1 - \frac{ж_2^{\,2}}{\mu^3}\left(\frac{l_P\, m_P}{q_P^{\,2}}\right)\right)$$

$G_0 = 7.74809172907834\ldots \times 10^{-5}\, sC^2/m^2kg$ predicted
$G_0 = 7.748091729\ldots \times 10^{-5}\, sC^2/m^2kg$ previously defined

Bohr electron radius

$$a_0 = \frac{1}{ж_1{}^2}\left(\frac{l_P\, m_P}{m_e}\right)\left(1 + ж_2{}^2\left(\frac{6}{Im(\rho_1)}\right)^{1/2}\left(\frac{l_P\, m_P}{q_P{}^2}\right)\right)$$

Where $Im(\rho_1)$ = the Imaginary part of the first nontrivial root of the zeta function ρ_1.

$a_0 = 5.29177210936601 \ldots \times 10^{-11}\, m$ predicted
$a_0 = 5.2917721092(17) \times 10^{-11}\, m$ measured

atomic mass constant

$$m_u = ж_1{}^2\, ж_2{}^2\,(\,q_P\, m_P\,)\sqrt{\frac{\zeta(3)}{b}}\left(1 + \left(\frac{8}{n}\right)\frac{Im(\,\rho_1\,)^2}{\Gamma(x_{min})}\left(\frac{l_P\, m_P}{q_P{}^2}\right)\right)$$

Where $\Gamma(x_{min})$ = the minimal value of the Gamma function for positive argument.

$m_u = 1.66053906659345 \ldots \times 10^{-27}\, kg$ predicted
$m_u = 1.66053906660(50) \times 10^{-27}\, kg$ measured

characteristic impedance

$$Z_0 = 4\pi\left(\frac{l_P{}^2 m_P}{t_P\, q_P{}^2}\right)\left(1 + \frac{2n^2}{b}\Gamma(x_{min})^4\left(\frac{l_P\, m_P}{q_P{}^2}\right)\right)$$

$Z_0 = 3.76730313668332 \ldots \times 10^2\ m^2kg/sC^2$ predicted
$Z_0 = 3.76730313668(57) \times 10^2\ m^2kg/sC^2$ measured

Stefan-Boltzmann constant

$$\sigma = \frac{\zeta(2)}{2n}\left(\frac{m_P}{t_P{}^3\, T_P{}^4}\right)\left(1+\left(\frac{P_{up}}{2\,ж_1}\right)^2\left(\frac{l_P\, m_P}{q_P{}^2}\right)\right)$$

Where P_{up} = the universal parabolic constant

$\sigma = 5.67037441935166 \ldots \times 10^{-8}\ kg/s^3K^4$ predicted
$\sigma = 5.670374419 \times 10^{-8}\ kg/s^3K^4$ previously defined

Avogadro constant

$$N_A = 6\left(\frac{ж_1{}^2\, e^{-\gamma}}{q_P\, m_P}\right)\left(1-\left(2\,ж_2\, P_{up}\right)^2\left(\frac{l_P\, m_P}{q_P{}^2}\right)\right)$$

$N_A = 6.02214076693260 \ldots \times 10^{23}\ 1/mol$ predicted
$N_A = 6.02214076 \times 10^{23}\ 1/mol$ previously defined

von Klitzing constant

$$R_K = \frac{2\pi}{ж_1{}^2}\left(\frac{l_P{}^2\, m_P}{t_P\, q_P{}^2}\right)\left(1+\frac{1}{2}Re\left(i^{i^{i^{\cdots}}}\right)ж_3ж_4\left(\frac{l_P\, m_P}{q_P{}^2}\right)\right)$$

$R_K = 2.5812807449400712 \ldots \times 10^4\ m^2kg/sC^2$ predicted
$R_K = 2.58128074434(84) \times 10^4\ m^2kg/sC^2$ measured

Hartree energy

$$E_h = ж_1{}^4\left(\frac{l_P{}^2\, m_e}{t_P{}^2}\right)\left(1-ж_2\left(3\,\frac{\omega_2{}^2}{n}\right)^2\left(\frac{l_P\, m_P}{q_P{}^2}\right)\right)$$

$E_h = 4.35974472221183 \ldots \times 10^{-18}\ m^2kg/s^2$ predicted
$E_h = 4.3597447222071(85) \times 10^{-18}\ m^2kg/s^2$ measured

the speed of light

$$c = \left(\frac{l_P}{t_P}\right)\left(1 - \left(\frac{3}{2}\right) j_{0,1} \left(\frac{l_P\, m_P}{q_P{}^2}\right)\right)$$

Where $j_{0,1}$ = the 1st root of the Bessel function

$c = 2.9979245814477 \ldots \times 10^8\, m/s$ predicted
$c = 2.99792458 \times 10^8\, m/s$ previously defined

spectral radiance constant

$$c_{1L} = 4\pi \left(\frac{l_P{}^4\, m_P}{t_P{}^3}\right)\left(1 - \sqrt{\frac{3}{2}}\, j_{0,1} \left(\frac{l_P\, m_P}{q_P{}^2}\right)\right)$$

$c_{1L} = 1.1910428690053 5\ldots \times 10^{-16}\ m^4 kg/s^3$ predicted
$c_{1L} = 1.191042869(53) \times 10^{-16}\ m^4 kg/s^3$ measured

classical electron radius

$$r_e = ж_1{}^2 \left(\frac{l_P\, m_P}{m_e}\right)\left(1 + n\,(\alpha_F - 1)\sqrt{\left(\frac{2}{3}\right) V_{fe}} \left(\frac{l_P\, m_P}{q_P{}^2}\right)\right)$$

$r_e = 2.81794032269510 \ldots \times 10^{-15}\, m$ predicted
$r_e = 2.8179403227(19) \times 10^{-15}\, m$ measured

Nuclear magneton

$$\mu_N = \frac{ж_1}{2}\left(\frac{l_P{}^2\, q_P\, m_P}{t_P\, m_+}\right)\left(1 + ж_3\, ж_4 \left(\frac{2}{3}\right)\frac{V_{fe}}{n}\left(\frac{l_P\, m_P}{q_P{}^2}\right)\right)$$

$\mu_N = 5.05078369897026 \ldots \times 10^{-27}\ m^2 C/s$ predicted
$\mu_N = 5.050783699(31) \times 10^{-27}\ m^2 C/s$ measured

2nd radiation constant

$$c_2 = 2\pi\, l_P T_P \left(1 - 2n\, e^{2\gamma}\, C_{CFP}{}^2 \left(\frac{l_P\, m_P}{q_P{}^2}\right)\right)$$

Where C_{CFP} = the real fixed point of the hyperbolic cotangent

$c_2 = 1.43877687708189 \ldots \times 10^{-2}\, m\, K$ predicted
$c_2 = 1.438776877 \ldots \times 10^{-2}\, m\, K$ previously defined

muon g-factor

$$g_\mu = -\frac{1}{4}(\alpha_F - 1)\sqrt{\frac{V_{fe}{}^b}{n}}\left(1 + \frac{1}{b}\sqrt{\frac{n}{b}}\left(\frac{l_P\, m_P}{q_P{}^2}\right)\right)$$

$g_\mu = -2.002331841799530 \ldots$ predicted
$g_\mu = -2.0023318418(13)$ measured

electron g-factor

$$g_e = -\frac{1}{4}(\alpha_F - 1)\sqrt{\frac{V_{fe}{}^b}{n}}\left(1 - \left(\frac{ж_2{}^2}{ж_1{}^3}\right)\frac{1}{2n^2}\left(\frac{3}{F_{FR}{}^2}\right)^2\left(\frac{l_P\, m_P}{q_P{}^2}\right)\right)$$

Where F_{FR} = the Fransén-Robinson constant

$g_e = -2.0023193043623145 \ldots$ predicted
$g_e = -2.00231930436256(35)$ measured

molar gas constant

$$R = 6\left(\frac{e^{-\gamma}}{q_P\, T_P}\right)\left(ж_1 \frac{l_P}{t_P}\right)^2\left(1 - \left(\frac{ж_2{}^2}{ж_1{}^3}\right)\frac{!n}{2n}\left(\frac{3}{F_{FR}{}^3}\right)^3\left(\frac{l_P\, m_P}{q_P{}^2}\right)\right)$$

$R = 8.31446261831660 \ldots\ m^2 kg/s^2 K\, mol$ predicted
$R = 8.314462618\ m^2 kg/s^2 K\, mol$ previously defined

magnetic flux constant

$$\Phi_0 = \left(\frac{\pi}{ж_1}\right)\left(\frac{l_P{}^2\, m_P}{t_P\, q_P}\right)\left(1 + 8\, G_{Go}\left(\frac{l_P\, m_P}{q_P{}^2}\right)\right)$$

Where G_{Go} = the Gompertz constant

$\Phi_0 = 2.06783384804388 \ldots \times 10^{-15}\, m^2kg/sC$ predicted
$\Phi_0 = 2.067833848 \ldots \times 10^{-15}\, m^2kg/sC$ previously defined

quantum of circulation

$$q_c = \pi\left(\frac{l_P{}^2\, m_P}{t_P\, m_e}\right)\left(1 + 8\, C_C\left(\frac{l_P\, m_P}{q_P{}^2}\right)\right)$$

Where C_C = Cahen's constant

$q_c = 3.63694755171206 \ldots \times 10^{-4}\, m^2/s$ predicted
$q_c = 3.6369475516(11) \times 10^{-4}\, m^2/s$ measured

proton g-factor

$$g_+ = ж_2\,(\,2\,B_1 + 1\,)\left(1 + 3\, C_c{}^{-4}\left(\frac{l_P\, m_P}{q_P{}^2}\right)\right)$$

Where B_1 = Merten's constant

$g_+ = +5.58569468933155 \ldots$ predicted
$g_+ = +5.5856946893(16)$ measured

Rydberg constant

$$R_\infty = \frac{ж_1{}^4}{4\pi\, l_P}\left(\frac{m_e}{m_P}\right)\left(1 - 4\, L_2\, b^{2/3}\left(\frac{l_P\, m_P}{q_P{}^2}\right)\right)$$

Where L_2 = the 2nd Lemniscate constant

$R_\infty = 1.09737315685870 \ldots \times 10^7\ 1/m$ predicted
$R_\infty = 1.097373156 8539(55) \times 10^7\ 1/m$ measured

proton gyromagnetic ratio

$$\gamma_+ = ж_1 \frac{\sqrt{L}}{6}\left(\frac{m_P}{q_P}\right)\left(1 - (\,3\,ж_2\,)^2\, {C_{CFP}}^3 \left(\frac{l_P\, m_P}{{q_P}^2}\right)\right)$$

Where C_{CFP} = the real fixed point of the hyperbolic cotangent

$\gamma_+ = 2.67522187452882 \ldots \times 10^8\ s/kg\ C$ predicted
$\gamma_+ = 2.6752218744(11) \times 10^8\ s/kg\ C$ measured

Neutron magnetic moment

$$N_\mu = -\left(\frac{ж_1}{2}\right)\left(\frac{!\,n}{\Gamma(n) - 1}\right)\left(\frac{{l_P}^2 q_P\, m_P}{t_P\, m_+}\right)$$

$N_\mu = -9.66236357094966 \ldots \times 10^{-27} C\ m^2/s$ predicted
$N_\mu = -9.6623647(23) \times 10^{-27} C\ m^2/s$ measured

Faraday constant

$$F = N_A\, ж_1\, q_P\left(1 + \left(\frac{n}{!\,n}\right)(\,ж_1\, ж_2\,)^{-1}\left(\frac{l_P\, m_P}{{q_P}^2}\right)\right)$$

$F = 9.64853321242908 \ldots \times 10^4\ \ C/mol$ predicted
$F = 9.648533212 \ldots \times 10^4\ \ C/mol$ previously defined

neutron g-factor

$$g_N = \frac{!\,n}{ж_1}\left(\frac{2^n}{n^3\, {V_{fe}}^n}\right)$$

$g_N = -3.826085907980 \ldots$ predicted
$g_N = -3.82608545(90)$ measured

gravitational coupling

$$\alpha_G = \left(\frac{m_e}{m_P}\right)^2$$

$\alpha_G = 1.75182147492404 \ldots \times 10^{-45}$ predicted
$\alpha_G = 1.7518(21) \times 10^{-45}$ measured

Compton angular frequency

$$\omega_C = \left(\frac{m_e}{t_P\, m_P}\right)$$

$\omega_c = 7.76345031059602 \ldots \times 10^{20}\ 1/s$ dominant term
$\omega_c = 7.763441 \times 10^{20}\ 1/s$ measured

Schwinger magnetic induction

$$S_{mi} = \left(\frac{{m_e}^2}{ж_1\, t_P\, q_P\, m_P}\right)$$

$S_{mi} = 4.41401084454021 \ldots \times 10^9\ kg/sC$ predicted
$S_{mi} = 4.419 \times 10^9\ kg/sC$ measured

gravitational constant

$$G = \left(\frac{{l_P}^3}{m_P\, {t_P}^2}\right)$$

$G = 6.67434500300505 \ldots \times 10^{-11}\ m^3/s^2kg$ predicted
$G = 6.67384(80) \times 10^{-11}\ m^3/s^2kg$ measured

Boltzmann constant

$$k_B = \left(\frac{{l_P}^2\, m_P}{{t_P}^2\, T_P}\right)$$

$k_B = 1.38064248507658 \ldots \times 10^{-23}\ m^2kg/s^2K$ predicted
$k_B = 1.380649 \times 10^{-23}\ m^2kg/s^2K$ previously defined

proton radius

$$\frac{r_e}{r_+} = L\sqrt{\frac{3}{2}}$$

$r_+ = 8.7749356797017 \times 10^{-16}\ m$ predicted
$r_+ = 8.751(61) \times 10^{-16}\ m$ measured

We are now in possession of the complete partition balance of the hyperbolic figure eight knot, the constructive legend of physical reality.[4]

[4] Note that the ratio of the strength of the electromagnetic force compared to the strong nuclear force is equal to the first solution of the hyperbolic vortex equation squared. Square that ratio again and we get the ratio of the classical electron radius compared to the Bohr electron radius.

$$\frac{em\ force}{strong\ force} = Ж_1{}^2 \qquad \frac{r_e}{a_0} = Ж_1{}^4$$

Additionally, the nonrelativistic ground state wave function of the hydrogen atom is,

$$\psi(r) = (\pi a_0{}^3)^{-1/2}\, e^{-r/a_0} = \frac{a_0{}^{3/2}}{\Gamma\left(\frac{1}{2}\right)}\, e^{-r/a_0}$$

And the radius values of Helium and Lithium 2… are.

$$He^+ = \frac{a_0}{2} \qquad Li^{2+} = \frac{a_0}{3} \qquad \ldots$$

The partition parameters of the hyperbolic figure eight knot (complete)

$n = 5$ number of unique rotations
$b = 7$ break in scale symmetry

$\phi_0 = 5.39125836832313 \ldots + 2\pi i(k) \quad k \in \mathbb{Z}$ 0th external rotation
$\phi_1 = 1.61625918175645 \ldots + 2\pi i(k)$ 1st external rotation
$\phi_2 = 1.87554596713962 \ldots + 2\pi i(k)$ 2nd external rotation
$\phi_3 = 2.17642683817579 \ldots + 2\pi i(k)$ 3rd external rotation
$\phi_4 = 1.41678698590795 \ldots + 2\pi i(k)$ 4th external rotation

$t_P = 5.39125836832313 \ldots \times 10^{-44}\ s$ Planck time
$l_P = 1.61625918175645 \ldots \times 10^{-35}\ m$ Planck length
$q_P = 1.87554596713962 \ldots \times 10^{-18}\ C$ Planck charge
$m_P = 2.17642683817579 \ldots \times 10^{-8}\ kg$ Planck mass
$T_P = 1.41678698590795 \ldots \times 10^{32}\ K$ Planck temperature

$V_{fe} = 2.02988321281930 \ldots$ figure eight knot complement volume
$e = 2.71828182845904$ Euler's number
$\pi = 3.14159265358979 \ldots$ Archimedes' constant
$ж_1 = 0.0854245431533304 \ldots$ 1st hyperbolic vortex constant
$ж_2 = 3.66756753485499 \ldots$ 2nd hyperbolic vortex constant
$ж_3 = -1.87649603900417 \ldots + 4.06615262615972 \ldots i$ 3rd
$ж_4 = -1.87649603900417 \ldots - 4.06615262615972 \ldots i$ 4th
$\alpha_F = 2.50290787509589 \ldots$ alpha Fiegenbaum constant
$\delta_F = 4.66920160910299 \ldots$ delta Fiegenbaum constant
$\gamma = 0.577215664901532 \ldots$ Euler-Mascheroni constant
$\mu = 1.45136923488338 \ldots$ nontrivial zero of the logarithmic integral
$G_{Ga} = 0.834626841674073 \ldots$ Gauss's constant
$L = 2.622057554292119 \ldots$ Lemniscate constant
$L_1 = 1.31102877714605 \ldots$ 1st Lemniscate constant
$L_2 = 0.599070117367796 \ldots$ 2nd Lemniscate constant
$D_{Do} = 0.739085133215160 \ldots$ Dottie number
$\omega_1 = 0.764977018528596 \ldots + 1.32497062714087 \ldots i$ omega_1
$\omega_2 = 1.529954037057192 \ldots$ omega_2 constant
$P_{up} = 2.29558714939263 \ldots$ universal parabolic constant
$K_{\_1} = 1.745405662407346 \ldots$ Khinchin harmonic mean
$j_{0,1} = 2.40482555769577 \ldots$ 1st root of the Bessel function
$L_{LL} = 0.662743419349181 \ldots$ Laplace limit
$C_{CFP} = 1.19967864025773 \ldots$ real fixed point of the hyperbolic cotangent
$\rho_1 = 0.5 + 14.1314251417346 \ldots i$ 1st nontrivial zero of the zeta function

$m_e = 9.10938370161994 \ldots \times 10^{-31}\ kg$	electron mass
$m_+ = 1.67262192371195 \ldots \times 10^{-27}\ kg$	proton mass
$m_N = 1.67492749802284 \ldots \times 10^{-27}\ kg$	neutron mass
$m_c = 2.27188026398178 \ldots \times 10^{-27}\ kg$	charm quark mass
$m_d = 8.44242715614137 \ldots \times 10^{-30}\ kg$	down quark mass
$m_u = 3.81810683898335 \ldots \times 10^{-30}\ kg$	up quark mass
$m_s = 1.82501207639326 \ldots \times 10^{-28}\ kg$	strange quark mass
$m_b = 7.45149186313980 \ldots \times 10^{-27}\ kg$	beauty (bottom) quark mass
$m_t = 3.08390948667753 \ldots \times 10^{-25}\ kg$	truth (top) quark mass
$m_H = 2.23150010999262 \ldots \times 10^{-25}\ kg$	Higgs boson mass
$m_Z = 1.62556627846185 \ldots \times 10^{-25}\ kg$	Z boson mass
$m_W = 1.43263881046217 \ldots \times 10^{-25}\ kg$	W boson mass
$m_\tau = 3.16754001786349 \ldots \times 10^{-27}\ kg$	tau mass
$m_\mu = 1.88353162775445 \ldots \times 10^{-28}\ kg$	muon mass
$m_{\nu_\tau} = 7.63391385156818 \ldots \times 10^{-39}\ kg$	tau neutrino mass
$m_{\nu_\mu} = 1.61737751049693 \ldots \times 10^{-43}\ kg$	muon neutrino mass
$m_{\nu_e} = 1.07825167366462 \ldots \times 10^{-43}\ kg$	electron neutrino mass

$\alpha = 7.29735257295522 \ldots \times 10^{-3}$	fine-structure constant
$e = 1.60217657405973 \ldots \times 10^{-19}\ C$	electron charge
$\lambda_C = 2.4263102389394 \ldots \times 10^{-12}\ m$	Compton wavelength
$K_J = 4.83597848400467 \ldots \times 10^{14}\ sC/m^2kg$	Josephson constant
$\hbar = 1.05457172593010 \ldots \times 10^{-34}\ m^2kg/s$	Planck's constant
$\varepsilon_0 = 8.85418781308692 \ldots \times 10^{-12}\ s^2C^2/m^3kg$	electric constant
$\kappa = 8.98755179196986 \ldots \times 10^{9}\ m^3kg/s^2C^2$	Coulomb's constant
$H_C = 3.87404614816855 \times 10^{-5}\ C^2/m^2kg$	quantized Hall conductance
$\mu_B = 9.27400994938886 \ldots \times 10^{-24}\ m^2C/s$	Bohr magneton
$\mu_0 = 1.25663706143747 \ldots \times 10^{-6}\ mkg/C^2$	magnetic constant
$c_1 = 3.74177185217629 \ldots \times 10^{-16}\ m^4kg/s^3$	1st radiation constant
$\sigma_e = 6.65246159951664 \ldots \times 10^{-29}\ m^2$	electron Thomson x section
$G_0 = 7.74809172900024 \ldots \times 10^{-5}\ sC^2/m^2kg$	conductance quantum
$a_0 = 5.29177210936601 \ldots \times 10^{-11}\ m$	Bohr electron radius
$m_u = 1.66053906659345 \ldots \times 10^{-27}\ kg$	atomic mass constant
$Z_0 = 3.76730313668332 \ldots \times 10^{2}\ m^2kg/sC^2$	characteristic impedance
$\sigma = 5.67037441935166 \ldots \times 10^{-8}\ kg/s^3K^4$	Stefan-Boltzmann constant
$N_A = 6.02214076693260 \ldots \times 10^{23}\ 1/mol$	Avogadro constant
$R_K = 2.58128074494007 \ldots \times 10^{4}\ m^2kg/sC^2$	von Klitzing constant
$E_h = 4.35974472220674 \ldots \times 10^{-18}\ m^2kg/s^2$	Hartree energy
$c = 2.9979245814477 \ldots \times 10^{8}\ m/s$	speed of light
$c_{1L} = 1.19104286900535 \ldots \times 10^{-16}\ m^4kg/s^3$	spectral radiance
$r_e = 2.81794032269510 \ldots \times 10^{-15}\ m$	classical electron radius

$\mu_N = 5.05078369897026 \ldots \times 10^{-27}\ m^2C/s$ Nuclear magneton
$c_2 = 1.43877687708189 \ldots \times 10^{-2}\ m\ K$ 2nd radiation constant
$g_\mu = -2.002331841799530 \ldots$ muon g-factor
$g_e = -2.0023193043623145 \ldots$ electron g-factor
$R = 8.31446261831660 \ldots\ m^2kg/s^2K\ mol$ molar gas constant
$\Phi_0 = 2.06783384804388 \ldots \times 10^{-15}\ m^2kg/sC$ magnetic flux constant
$q_c = 3.63694755171206 \ldots \times 10^{-4}\ m^2/s$ quantum of circulation
$g_+ = +5.58569468933155 \ldots$ proton g-factor
$R_\infty = 1.09737315685870 \ldots \times 10^7\ 1/m$ Rydberg constant
$\gamma_+ = 2.67522187452882 \ldots \times 10^8\ s/kg\ C$ proton gyromagnetic ratio
$N_\mu = -9.66236357094966 \ldots \times 10^{-27} C\ m^2/s$ neutron magnetic moment
$F = 9.64853321242908 \ldots \times 10^4\ C/mol$ Faraday constant
$g_N = -3.826085907980 \ldots$ neutron g-factor
$\alpha_G = 1.75182147492404 \ldots \times 10^{-45}$ gravitational coupling constant
$\omega_c = 7.76345031059602 \ldots \times 10^{20}\ 1/s$ Compton angular frequency
$S_{mi} = 4.41401084454021 \ldots \times 10^9\ kg/sC$ Schwinger magnetic induction
$G = 6.67434500300505 \ldots \times 10^{-11}\ m^3/s^2kg$ gravitational constant
$k_B = 1.38064248507658 \ldots \times 10^{-23}\ m^2kg/s^2K$ Boltzmann constant

$F_{FR} = 2.8077702420285 \ldots$ Fransén-Robinson constant
$x_{min} = 1.461632144968362 \ldots$ min Γ function value for positive argument
$\zeta(3) = 1.20205690315959 \ldots$ Aprey's constant
$\zeta(2) = \pi^2/6$ zeta of 2
$i^{i^{i\cdots}} = 0.438282936727032 \ldots + 0.360592471871385 \ldots i$ *i* power tower
$G_{Go} = 0.596347362323194 \ldots$ Gompertz constant
$C_C = 0.643410546288338 \ldots$ Cahen's constant
$B_1 = 0.261497212847642 \ldots$ Merten's constant
$B_2 = 1.9021605831040$ Brun's constant

Where the black digits represent previously known values (either measured or geometrically known), and the green digits represent predictions.

number of digits measured needing explanation = 436
number of predicted digits verified by measurement = 436
number of additionally predicted digits = 664
total prediction digits = 1100

Chapter 11—notice

"[F]or nearly a century, physicists have been warned against trying to understand physics... The standard mantra, as has been repeated over and over, is "Shut Up And Calculate". That is, simply do not think about how to understand the world: just grind out predictions. And not being trained or encouraged—indeed being actively discouraged—from trying to think clearly about that question leaves most physicists in no position to make well-considered comments about it."

Tim Maudlin

"We lost academic freedom. And then we ushered in peer review, which was a disaster, and then we lost funding."

Eric Weinstein

A common response I hear is "I can't wait to hear what other physicists have to say about this". Somehow people are blissfully unaware of the situation we are in. There are no people professionally in this conversation. Period. No physicists, or philosophers, or mathematicians in the entire world are currently getting paid (including me) to work on this topic. Not a penny. This conversation is up to *us* to resurrect.

When the average person thinks of *physics* they seem to think of it as it was in the time of Einstein, when researchers could still professionally pursue questions of their own choosing. Today, things are different.

Under the modern peer review system, those controlling the research money control the conversation of humanity; deciding what get's explored and, more importantly, what doesn't. And, as every physics graduate student discovers with shock, those in charge have deemed the *theory of everything* an *unfundable* topic—the modern way of saying "forbidden conversation".

In today's world, if you want to get paid, then you must pick a different topic. And with that slight-of-hand the most honest questions we can ask are no longer allowed. The modern focus has backfired on us all, effectively snuffing out the most noble part of existence—honest exploration of one's reality.

I believe we have just outlined the most accurate understanding of reality's structure that has ever existed, and that this elegant perspective is capable of initiating the next great awakening of humanity. But seeing how this is happening at a point in history in which the battle has already been lost, in which there's nobody left in the conversation, the first step it to make this a world in which this conversation is, once again, safe.

To date, despite having been trained as a physicist, and having worked at NASA, I have yet to meet another individual that I know has taken the first step in figuring out the biggest puzzle of all time—compiling a list of the parameters that define the mysteries of quantum field theory. Literally everybody, no matter what their position or financial fortune, finds themselves in a world in which it is not safe to ask the most honest questions one can ask about the make-up of reality.

If you'd like to live in a world that incentivizes its individuals to think about how to understand the world, *please* consider supporting this research.

Chapter 12—coming up

One of the delightful surprises of this construction is that one of its partition parameters inversely depends on the square root of the Imaginary part of the first non-trivial zero of the Riemann zeta function ($Im(\rho_1)$), while another depends on the square of that number—structurally balancing the first zero in the most famous sequence of zeros in all of math.

Obviously, there's more story to tell here, but I've been told that every math equation put into a book cuts its audience in half. So, since, by that logic it should take quite some time to find a reader all the way down here, I'll cut the story off here for now and get back to working on enhancing my own understanding.

If you would like to join this effort to understand the world, I recommend getting the extended version of this book titled "44". It includes several appendices of organized geometric relations, all of which are consequences of this partition balance—relaying secrets about reality's construction never before seen. I also recommend verifying everything in this book for yourself and really getting to know the geometric equations that define the balances of reality—the logic of persistence.

Appendix a: mass (equations only)

$$m_e = 2V_{fe}\, m_P{}^4\left(1+\left(\sinh\left(\sinh\left(\frac{!\,\boldsymbol{n}}{\boldsymbol{b}}\right)\right)^{-1}\right)^{-1}\left(\frac{l_P\, m_P}{q_P{}^2}\right)\right)$$

$$\left(\frac{m_e}{m_+}\right)\left(\frac{ж_2}{ж_1}\right)^2 = \left(\frac{2}{3}\right)^2 (\alpha_F - 1)^2\left(1-\left(\frac{1}{3}\right)e^{3\gamma}\left(\frac{l_P\, m_P}{q_P{}^2}\right)\right)$$

$$\left(\frac{m_N - m_+}{m_e}\right) = \left(\frac{1}{3}\right)(\mu + 3 + \pi)\left(1+\left(\frac{2}{3}\right)e^{3\gamma}\left(\frac{l_P\, m_P}{q_P{}^2}\right)\right)$$

- -

$$\left(\frac{m_H - m_Z}{m_W}\right) = \left(\frac{1}{3}\right)^2 (\mu + 3 + \pi)(2^{-1})\left(1+\left(\frac{ж_2{}^2}{ж_1{}^3}\right)G_{Ga}{}^{-3}\left(\frac{\boldsymbol{n}}{\boldsymbol{b}}\right)\left(\frac{l_P\, m_P}{q_P{}^2}\right)\right)$$

$$\left(\frac{m_N - m_+}{m_e}\right) = \left(\frac{1}{3}\right)(\mu + 3 + \pi)(2^0)\left(1+\left(\frac{2}{3}\right)e^{3\gamma}\left(\frac{l_P\, m_P}{q_P{}^2}\right)\right)$$

$$\left(\frac{m_d - m_u}{m_e}\right) = \left(\frac{1}{3}\right)(\mu + 3 + \pi)(2^{+1})\left(1+\left(\frac{ж_2{}^2}{ж_1{}^3}\right)D_{Do}{}^3\sqrt{\frac{2}{\boldsymbol{b}}\left(\frac{l_P\, m_P}{q_P{}^2}\right)}\right)$$

external phases

- -

internal folds

$$\left(\frac{m_c - 2m_s}{m_\mu}\right) = \left(\frac{1}{3}\right)(\mu + 3 + \pi)(2^2)$$

$$\left(\frac{m_t - m_\tau}{m_b}\right) = \left(\frac{1}{3}\right)(\mu + 3 + \pi)\left(2^{2^2}\right)\left(1+\boldsymbol{b}\,\frac{\sqrt{Re(\omega_1)}}{ж_1{}^4}\left(\frac{l_P\, m_P}{q_P{}^2}\right)\right)$$

$$\left(\frac{m_{\nu_\tau} - m_{\nu_\mu}}{m_{\nu_e}}\right) = \left(\frac{1}{3}\right)(\mu + 3 + \pi)\left(2^{2^{2^2}}\right)$$

$$\frac{m_b + m_c + m_t}{\left(\sqrt{m_b} + \sqrt{m_c} + \sqrt{m_t}\right)^2} = \left(\frac{2}{3}\right)^3 (\alpha_F - 1)^2 \left(1 + \left(\frac{ж_2{}^3}{ж_1{}^2}\right) \frac{V_{fe}{}^4}{2^n} \left(\frac{l_P\, m_P}{q_P{}^2}\right)\right)$$

$$\left(\frac{m_e}{m_+}\right)\left(\frac{ж_2}{ж_1}\right)^2 = \left(\frac{2}{3}\right)^2 (\alpha_F - 1)^2 \left(1 - \left(\frac{1}{3}\right) e^{3\gamma} \left(\frac{l_P\, m_P}{q_P{}^2}\right)\right)$$

$$\frac{m_e + m_\mu + m_\tau}{\left(\sqrt{m_e} + \sqrt{m_\mu} + \sqrt{m_\tau}\right)^2} = \left(\frac{2}{3}\right)^1 (\alpha_F - 1)^0 \left(1 - (3\, ж_2{}^2)\, P_{up} \left(\frac{l_P\, m_P}{q_P{}^2}\right)\right)$$

twist-bifurcation balance

- -

flip-split balance

$$\frac{m_H + m_Z + m_W}{\left(\sqrt{m_H} + \sqrt{m_Z} + \sqrt{m_W}\right)^2} = \left(\frac{2}{3}\right)^{-\frac{1}{2}} (\alpha_F - 1)^{-e^{2\gamma}} \left(1 + \left(\frac{ж_2}{ж_1}\right)^2 \left(2nb\, (V_{fe})^{\frac{1}{3}}\right)^{-\frac{1}{2}} \left(\frac{l_P\, m_P}{q_P{}^2}\right)\right)$$

$$\frac{m_u + m_s + m_d}{\left(\sqrt{m_u} + \sqrt{m_s} + \sqrt{m_d}\right)^2} = \gamma$$

$$\frac{m_{\nu_e} + m_{\nu_\mu} + m_{\nu_\tau}}{\left(\sqrt{m_{\nu_e}} + \sqrt{m_{\nu_\mu}} + \sqrt{m_{\nu_\tau}}\right)^2} = \left(\frac{2}{3}\right)^{-\frac{1}{2}} (\alpha_F - 1)^{+e^{\pi i}} \left(1 - \left(\frac{ж_2}{ж_1}\right)^2 \left(2nb\, (V_{fe})^{\frac{2}{3}}\right)^{-\frac{1}{2}} \left(\frac{l_P\, m_P}{q_P{}^2}\right)\right)$$

- -

$$\frac{m_e}{m_c}\left(\frac{ж_2}{ж_1}\right)^2 = D_{DO}$$

$$\frac{m_e}{m_d} = \frac{\mu}{ж_2{}^2}$$

$$\frac{m_e}{m_u} = \frac{2}{\omega_2{}^n}$$

$$\frac{m_e}{m_s}\left(\frac{!n}{\sqrt{n}}\right) = \frac{\sqrt{K_{\_1}}}{ж_2{}^2}$$

$$\frac{m_e}{m_W}\left(\frac{ж_2{}^2}{ж_1{}^3}\right) = (Im(\rho_1))^{-3/4}$$

$$\frac{m_e}{m_b}\left(\frac{ж_2}{ж_1}\right)^2 = \frac{L_1{}^3}{2n}$$

$$\frac{m_e}{m_t}\left(\frac{ж_2{}^3}{ж_1{}^2}\right) = \frac{L_2}{6n}$$

$$\frac{m_e}{m_H}\left(\frac{ж_2}{ж_1}\right)^2 = \frac{b}{2\, j_{0,1}{}^b}$$

$$\frac{m_e}{m_Z}\left(\frac{ж_2{}^3}{ж_1{}^2}\right) = \left(\frac{2}{3}\right)^3 L_{LL}{}^n$$

$$\frac{m_e}{m_\tau}\left(\frac{ж_2}{ж_1}\right)^2 = \left(\frac{3}{{V_{fe}}^2}\right)^2 \left(1 - {ж_2}^2 \operatorname{sech}\left(\tan\left(\frac{1}{2}\right)\right)\left(\frac{l_P\, m_P}{{q_P}^2}\right)\right)$$

$$\frac{m_e}{m_\mu} = {ж_1}^2\, L_{LL} \left(1 + \frac{\boldsymbol{b}\, ж_2}{{V_{fe}}^2}\left(\frac{!\boldsymbol{n}}{\sqrt{\boldsymbol{n}}}\right)\left(\frac{l_P\, m_P}{{q_P}^2}\right)\right)$$

$$\frac{l_P}{m_{\nu_\tau}}\left(\frac{ж_1}{ж_2}\right)^2 = 2^{1/\boldsymbol{n}}\left(1 - (\sqrt{2} - 1)\left(\frac{ж_2}{ж_1}\right)^2\left(\frac{l_P\, m_P}{{q_P}^2}\right)\right)$$

$$m_{\nu_e} = 3\, ф_0(2\boldsymbol{n})^{-!\boldsymbol{n}}\, kg$$

$$m_{\nu_e} = 2\, ф_0(2\boldsymbol{n})^{-!\boldsymbol{n}}\, kg$$

Appendix b: constants of Nature (equations only)

$$\alpha = ж_1{}^2$$

$$e = ж_1 q_P$$

$$\lambda_C = 2\pi\left(\frac{l_P\, m_P}{m_e}\right)\left(1 + \alpha_F\left(\frac{\boldsymbol{b}}{2}\right)\left(\frac{l_P\, m_P}{q_P{}^2}\right)\right)$$

$$K_J = \frac{ж_1}{\pi}\left(\frac{t_P\, q_P}{l_P{}^2\, m_P}\right)\left(1 - \left(\frac{e^{2\gamma}}{Im(\omega_1)^3}\right)\left(\frac{\boldsymbol{b}}{2}\right)\left(\frac{l_P\, m_P}{q_P{}^2}\right)\right)$$

$$\hbar = \left(\frac{l_P{}^2\, m_P}{t_P}\right)\left(1 + \sqrt{\frac{\boldsymbol{b}\pi}{\zeta(3)}}\left(\frac{l_P\, m_P}{q_P{}^2}\right)\right)$$

$$\varepsilon_0 = \frac{1}{4\pi}\left(\frac{t_P{}^2\, q_P{}^2}{l_P{}^3\, m_P}\right)\left(1 - \left(\frac{\boldsymbol{n}^2}{2^{\boldsymbol{n}}}\right)\left(\frac{l_P\, m_P}{q_P{}^2}\right)\right)$$

$$\kappa = \left(\frac{l_P{}^3\, m_P}{t_P{}^2\, q_P{}^2}\right)\left(1 + \left(\frac{\boldsymbol{n}^2}{2^{\boldsymbol{n}}}\right)\left(\frac{l_P\, m_P}{q_P{}^2}\right)\right)$$

$$H_C = \frac{ж_1{}^2}{2\pi}\left(\frac{t_P\, q_P{}^2}{l_P{}^2 m_P}\right)\left(1 - ж_2\left(\frac{l_P\, m_P}{q_P{}^2}\right)\right)$$

$$\mu_B = \frac{ж_1}{2}\left(\frac{l_P{}^2\, q_P\, m_P}{t_P\, m_e}\right)\left(1 + \frac{2\pi^2}{ж_2}\left(\frac{l_P\, m_P}{q_P{}^2}\right)\right)$$

$$\mu_0 = 4\pi\left(\frac{l_P\, m_P}{q_P{}^2}\right)\left(1 + ж_2\, e^{\gamma}\sqrt{\frac{3}{2}}\left(\frac{l_P\, m_P}{q_P{}^2}\right)\right)$$

$$c_1 = 4\pi^2\left(\frac{l_P{}^4\, m_P}{t_P{}^3}\right)\left(1 - ж_2{}^2(\,2\gamma - 1\,)\left(\frac{l_P\, m_P}{q_P{}^2}\right)\right)$$

$$\sigma_e = \left(\frac{2}{3}\right)4\pi\, ж_1{}^4\left(\frac{l_P{}^2\, m_P}{m_e}\right)^2\left(1 + ж_2{}^2(\,L - 1\,)\left(\frac{l_P\, m_P}{q_P{}^2}\right)\right)$$

$$G_0 = \frac{ж_1{}^2}{\pi}\left(\frac{t_P\, q_P{}^2}{l_P{}^2 m_P}\right)\left(1 - \frac{ж_2{}^2}{\mu^3}\left(\frac{l_P\, m_P}{q_P{}^2}\right)\right)$$

$$a_0 = \frac{1}{ж_1{}^2}\left(\frac{l_P\, m_P}{m_e}\right)\left(1 + ж_2{}^2\left(\frac{6}{Im(\rho_1)}\right)^{1/2}\left(\frac{l_P\, m_P}{q_P{}^2}\right)\right)$$

$$m_u = ж_1{}^2\, ж_2{}^2\, (\, q_P\, m_P\,)\sqrt{\frac{\zeta(3)}{\boldsymbol{b}}}\left(1 + \left(\frac{8}{\boldsymbol{n}}\right)\frac{Im(\,\rho_1\,)^2}{\Gamma(x_{min})}\left(\frac{l_P\, m_P}{q_P{}^2}\right)\right)$$

$$Z_0 = 4\pi\left(\frac{l_P{}^2 m_P}{t_P\, q_P{}^2}\right)\left(1 + \frac{2\boldsymbol{n}^2}{\boldsymbol{b}}\Gamma(x_{min})^4\left(\frac{l_P\, m_P}{q_P{}^2}\right)\right)$$

$$\sigma = \frac{\zeta(2)}{2\boldsymbol{n}}\left(\frac{m_P}{t_P{}^3\, T_P{}^4}\right)\left(1 + \left(\frac{P_{up}}{2\, ж_1}\right)^2\left(\frac{l_P\, m_P}{q_P{}^2}\right)\right)$$

$$N_A = 6\left(\frac{ж_1{}^2\, e^{-\gamma}}{q_P\, m_P}\right)\left(1 - (\, 2\, ж_2\, P_{up}\,)^2\left(\frac{l_P\, m_P}{q_P{}^2}\right)\right)$$

$$R_K = \frac{2\pi}{ж_1{}^2}\left(\frac{l_P{}^2\, m_P}{t_P\, q_P{}^2}\right)\left(1 + \frac{1}{2}Re\left(i^{i^{\cdots}}\right)ж_3 ж_4\left(\frac{l_P\, m_P}{q_P{}^2}\right)\right)$$

$$E_h = ж_1{}^4\left(\frac{l_P{}^2\, m_e}{t_P{}^2}\right)\left(1 - ж_2\left(3\,\frac{\omega_2{}^2}{\boldsymbol{n}}\right)^2\left(\frac{l_P\, m_P}{q_P{}^2}\right)\right)$$

$$c = \left(\frac{l_P}{t_P}\right)\left(1 - \left(\frac{3}{2}\right) j_{0,1}\left(\frac{l_P\, m_P}{q_P{}^2}\right)\right)$$

$$c_{1L} = 4\pi\left(\frac{l_P{}^4\, m_P}{t_P{}^3}\right)\left(1 - \sqrt{\frac{3}{2}}\, j_{0,1}\left(\frac{l_P\, m_P}{q_P{}^2}\right)\right)$$

$$r_e = ж_1{}^2 \left(\frac{l_P\, m_P}{m_e} \right) \left(1 + n\,(\alpha_F - 1) \sqrt{\left(\frac{2}{3}\right) V_{fe}} \left(\frac{l_P\, m_P}{q_P{}^2} \right) \right)$$

$$\mu_N = \frac{ж_1}{2} \left(\frac{l_P{}^2\, q_P\, m_P}{t_P\, m_+} \right) \left(1 + ж_3\, ж_4 \left(\frac{2}{3}\right) \frac{V_{fe}}{n} \left(\frac{l_P\, m_P}{q_P{}^2} \right) \right)$$

$$c_2 = 2\pi\, l_P T_P \left(1 - 2n\, e^{2\gamma}\, C_{CFP}{}^2 \left(\frac{l_P\, m_P}{q_P{}^2} \right) \right)$$

$$g_\mu = -\frac{1}{4}(\alpha_F - 1) \sqrt{\frac{V_{fe}{}^b}{n}} \left(1 + \frac{1}{b} \sqrt{\frac{n}{b}} \left(\frac{l_P\, m_P}{q_P{}^2} \right) \right)$$

$$g_e = -\frac{1}{4}(\alpha_F - 1) \sqrt{\frac{V_{fe}{}^b}{n}} \left(1 - \left(\frac{ж_2{}^2}{ж_1{}^3} \right) \frac{1}{2n^2} \left(\frac{3}{F_{FR}{}^2} \right)^2 \left(\frac{l_P\, m_P}{q_P{}^2} \right) \right)$$

$$R = 6 \left(\frac{e^{-\gamma}}{q_P\, T_P} \right) \left(ж_1 \frac{l_P}{t_P} \right)^2 \left(1 - \left(\frac{ж_2{}^2}{ж_1{}^3} \right) \frac{!\,n}{2n} \left(\frac{3}{F_{FR}{}^3} \right)^3 \left(\frac{l_P\, m_P}{q_P{}^2} \right) \right)$$

$$\Phi_0 = \left(\frac{\pi}{ж_1} \right) \left(\frac{l_P{}^2\, m_P}{t_P\, q_P} \right) \left(1 + 8\, G_{Go} \left(\frac{l_P\, m_P}{q_P{}^2} \right) \right)$$

$$q_c = \pi \left(\frac{l_P{}^2\, m_P}{t_P\, m_e} \right) \left(1 + 8\, C_C \left(\frac{l_P\, m_P}{q_P{}^2} \right) \right)$$

$$g_+ = ж_2\,(2\,B_1 + 1) \left(1 + 3\, C_c{}^{-4} \left(\frac{l_P\, m_P}{q_P{}^2} \right) \right)$$

$$R_\infty = \frac{ж_1{}^4}{4\pi\, l_P} \left(\frac{m_e}{m_P} \right) \left(1 - 4\, L_2\, b^{2/3} \left(\frac{l_P\, m_P}{q_P{}^2} \right) \right)$$

$$\gamma_+ = ж_1 \frac{\sqrt{L}}{6} \left(\frac{m_P}{q_P} \right) \left(1 - (3\, ж_2)^2\, C_{CFP}{}^3 \left(\frac{l_P\, m_P}{q_P{}^2} \right) \right)$$

$$N_\mu = -\left(\frac{ж_1}{2}\right)\left(\frac{!\,\boldsymbol{n}}{\Gamma(\boldsymbol{n})-1}\right)\left(\frac{{l_P}^2 q_P\, m_P}{t_P\, m_+}\right)$$

$$F = N_A\, ж_1\, q_P \left(1+\left(\frac{\boldsymbol{n}}{!\,\boldsymbol{n}}\right)(ж_1\, ж_2)^{-1}\left(\frac{l_P\, m_P}{{q_P}^2}\right)\right)$$

$$g_N = \frac{!\,\boldsymbol{n}}{ж_1}\left(\frac{2^{\boldsymbol{n}}}{\boldsymbol{n}^3\, {V_{fe}}^{\boldsymbol{n}}}\right)$$

$$\alpha_G = \left(\frac{m_e}{m_P}\right)^2$$

$$\omega_C = \left(\frac{m_e}{t_P\, m_P}\right)$$

$$S_{mi} = \left(\frac{{m_e}^2}{ж_1\, t_P\, q_P\, m_P}\right)$$

$$G = \left(\frac{{l_P}^3}{m_P\, {t_P}^2}\right)$$

$$k_B = \left(\frac{{l_P}^2\, m_P}{{t_P}^2\, T_P}\right)$$

$$\frac{r_e}{r_+} = L\sqrt{\frac{3}{2}}$$

$$\frac{em\ force}{strong\ force} = {ж_1}^2 \qquad\qquad \frac{r_e}{a_0} = {ж_1}^4$$

$$\psi(r) = (\pi {a_0}^3)^{-1/2}\, e^{-r/a_0} = \frac{{a_0}^{3/2}}{\Gamma\left(\frac{1}{2}\right)}\, e^{-r/a_0}$$

$$He^+ = \frac{a_0}{2} \qquad\qquad Li^{2+} = \frac{a_0}{3} \qquad\qquad \ldots$$

Appendix c: geometric identities

$\sum \phi_k = \phi_0 + \phi_1 + \phi_2 + \phi_3 + \phi_4$ sum of rotations

$\prod \phi_k = \phi_0\ \phi_1\ \phi_2\ \phi_3\ \phi_4$ product of rotations

$\sum d_k = d_0 + d_1 + d_2 + d_3 + d_4 = 4^2 + 11^2 = 137$ sum of powers

geometry	**circumference of a circle**	**area of a circle**
spherical	$2\pi \sin(r)$	$2\pi(1 - \cos(r))$
Euclidean	$2\pi(r)$	πr^2
hyperbolic	$2\pi \sinh(r)$	$2\pi(\cosh(r) - 1)$

geometry	**Pythagorean theorem**
spherical	$\cos(a)\cos(b) = \cos(c)$
Euclidean	$a^2 + b^2 + c^2$
Hyperbolic	$\cosh(a)\cosh(b) = \cosh(c)$

geometry	**# of parallels**	$\sum \boldsymbol{\theta}$ ***in*** Δ	**cir/diameter**	**curvature**
spherical	∞	< 180°	$> \pi$	< 0
Euclidean	1	180°	π	0
Hyperbolic	0	> 180°	$< \pi$	> 0

fundamental action	**fixed point of fundamental action**	
$\cos(x) = x$	D_{DO}	the Dottie number
$\coth(x) = x$	C_{CFP}	Real fixed point of the hyperbolic cotangent
$\sinh(x) = x$	0	
$\sinh^{-1}(x) = x$	0	
$\tanh(x) = x$	0	
$\tanh^{-1}(x) = x$	0	
$\sin(x) = x$	0	
$\sin^{-1}(x) = x$	0	

$$\cos^2(x) + \sin^2(x) = +1$$ *circle*
$$-\cosh^2(x) + \sinh^2(x) = -1$$ *hyperbola*

$$x^2 + y^2 = +1$$ *circle*
$$-x^2 + y^2 = -1$$ *hyperbola*

$$(x^2 + y^2)^2 = -(-x^2 + y^2)$$ lemniscate
$$(circle)^2 = -hyperbola$$ lemniscate

$$r^2 = 2a^2 \cos(2\theta)$$ lemniscate

$$\pm(\sin^2(x) + \cos^2(x) + \sinh^2(x) - \cosh^2(x)) = \pm 0$$
$$\pm(\sin^2(x) + \cos^2(x) - \sinh^2(x) + \cosh^2(x)) = \pm 2$$

$$\pm(\text{circle} + \text{hyperbola}) = \pm 0$$
$$\pm(\text{circle} - \text{hyperbola}) = \pm 2$$

$$sinh(x) = \frac{1}{2}(e^x - e^{-x})$$ hyperbolic sine

$$cosh(x) = \frac{1}{2}(e^x + e^{-x})$$ hyperbolic cosine

$$\cos^2(x) = \cosh^2(i\,x)$$

$$-e^{-x} = \sinh(x) - \cosh(x)$$ rotation of hyperbolic union

$$e^{ix} = \cos(x) + i\sin(x)$$ rotation of circular union

$$2\cosh(\log x) = \frac{1}{x} + x = e^{\log x} + e^{-\log x}$$

$$\frac{1}{2}\left(\left(\frac{1}{2}(e^{-x} + e^x)\right) + 1\right) = \left(\cosh\left(\frac{x}{2}\right)\right)^2$$

$$\frac{1}{2}(1 - x^2) + \frac{1}{2}(1 + x^2) = 1$$

$$\operatorname{sech}(x) = \frac{1}{\cosh(x)} = \frac{2}{e^{-x} + e^{x}} = \frac{2\cosh(x)}{\cosh(2x) + 1}$$

$$n^2 - \Gamma(n) = 1$$

$$\cos(m\,\pi) = (-1)^m \qquad m \in \mathbb{Z}$$

$$b\cosh(\ln b) = n^2 \qquad \text{hyperbolic exponential square balance}$$

$$b\sinh(\ln b) = \Gamma(n) \qquad \text{hyperbolic boundary factorization}$$

$$b - 1 = n + 1 \qquad \text{hyperbolic} - \text{circular phase balance}$$

$$\cosh(\ln b) = \frac{n^2}{b} \qquad \text{hyperbolic} - \log - \text{square balance}$$

$$\frac{1}{2}(b^2 + 1) = n^2 \qquad \text{square split balance}$$

$$\sqrt{2n^2 - 1} = b$$

$$\sum_{n=0}^{\infty} (2n)!\cosh(x) = \sum_{n=0}^{\infty} x^{2n} \qquad \text{hyperbolic double cover factorization}$$

$$(\log 2)^2 = Li_1(-1)^2 = 4\coth^{-1}(3)^2$$

$$\sinh^{-1}\left(\frac{1}{2}\right) = \log\phi$$

$$\operatorname{csch}^{-1}(2) = \log\phi \qquad \phi = \text{the golden ratio}$$

$$\coth^{-1}(0) = -\frac{\pi}{2}i$$
$$\coth^{-1}(\infty) = 0$$
$$\coth^{-1}(1) = \infty$$
$$\coth^{-1}(i) = -\frac{\pi}{4}i$$

$i = \sqrt{-1}$ the complex root

$i^{i^2} = -i$ complex inversion power squaring

$\left(i^{i^2}\right)^2 = -1$ Möbius power squaring

$\left(\left(i^{i^2}\right)^2\right)^2 = 1$ power squaring completion

$\left(i^i\right)^2 = i^{2i} = e^{-\pi}$ square balance of internal split rotation

$\left(\left(i^i\right)^2\right)^2 = i^{4i} = e^{-2\pi}$ four fold division of circular projection

$\left(\left(\left(i^i\right)^2\right)^2\right)^2 = i^{8i} = e^{-4\pi}$ eight fold division of spherical projection

$i^i = e^{-\frac{\pi}{2}}$

$e^{\pi} = (-1)^{-i}$ $\sqrt{e^{\pi i}} = i$

$$i^{i^{i\cdots}} = -\frac{W(-\ln(i))}{\ln(i)} = \left(\frac{2}{\pi}i\right) W\left(-\frac{\pi}{2}i\right)$$

$$W(x)\, e^{W(x)} = x$$

$$W(x) = \frac{1}{\pi} Re \int_0^{\pi} \ln\left(\frac{e^{e^{it}} - xe^{-it}}{e^{e^{it}} - xe^{it}}\right) dt$$

$$\int_0^e W_0(x)\, dx = e - 1$$

$$\int_0^{\infty} \frac{dx}{(e^x - x^2)^2 + \pi^2} = \frac{1}{1 + W(1)}$$

$$e^x = \frac{x^0}{0!} + \frac{x^1}{1!} + \frac{x^2}{2!} + \cdots$$

$$\sum_{k=0}^{n-1} e^{2\pi i\left(\frac{k}{n}\right)} = 0$$ $n = 2$ gives Euler's identity

$$e^{ix} = \cos(x) + i\sin(x)$$

$$e^{\frac{1}{\sqrt{3}}(i \pm j \pm k)\pi} + 1 = 0$$ quaternion

$$e^{(a_1 i_1 + a_2 i_2 + \cdots + a_7 i_7)\pi} + 1 = 0$$ octonion

$$Ei(\log x) = li(x)$$ $li(e^x) = Ei(x)$

$$\frac{x!}{!x} = e\,\frac{\Gamma(x+1)}{\Gamma(x+1,-1)}$$

$\Gamma(x+1) = \Gamma(x-1)$, when $x = \frac{1}{2} \pm \frac{\sqrt{n}}{2}$

$$\frac{\pi}{8} = \int_0^1 \sqrt{x(1-x)}\; dx$$

$$1 = \int_0^e \frac{1}{x}\, dx$$

$$x^t = e^{\log(x)\,t}$$

$$\sqrt{\pi} = \Gamma\left(\frac{1}{2}\right) = \left(-\frac{1}{2}\right)!$$

$$\pi = \sqrt{6\,\zeta(2)}$$

$$\zeta(-k) = -\frac{B_{k+1}}{k+1}$$ Zeta – Bernoulli connection

$$\zeta(-2k) = 0$$ trivial zeros of the Riemann zeta function

$\zeta(0) = -\frac{1}{2}$ the Riemann zeta balance

$b\,\zeta(3) = \zeta\left(3, \frac{1}{2}\right)$ Aprey's split

$\sqrt{\left(\frac{3}{2}\right)\zeta(2)} = \frac{\pi}{2}$ $\left(\frac{3}{2}\right)\zeta(2) = \frac{\pi^2}{4}$ $\zeta(2) = \frac{\pi^2}{6}$

$\prod_{k=1}^{\infty}(1 - x^k)^{\mu(k)/k} = e^{-x}$ $\mu(k) =$ the Mertens function

$\sum_{k=1}^{\infty}\frac{\mu(k)}{k^s} = \frac{1}{\zeta(s)}$ Mertens inverse zeta summation

Acknowledgements

I thank George Cassiday for the passion of his ETI class.

I thank Albert Einstein for teaching me true courage of aim, Leonhard Euler for his insightful reach, and Gene Shoemaker (the man on the moon) for his bold example.

I thank my Num for pinky holds, Shangri-La, and for really meaning it.

I thank my $\sqrt{-1}$ Cloud for her delightful imagination.

I thank Elaine, Phil, and Matt Emmi for the way they gave me a key.

I thank David Heggli for believing that maintaining an open-minded discussion, for however long it takes, is what friends do.

I thank Jeff Chapple—for all the Alfred stuff, and David Cantu for what may or may not have been said in the cone of silence.

I thank Anaximander, Michael Faraday, Benoit Mandelbrot, Louis de Broglie, David Bohm, Stephen Wolfram, Shelly Goldstein, Detlef Dürr, Nino Zanghí, Grant Sanderson (3Blue1Brown), Derek Muller (Veritasium), Brady Haran (Numberphile), Franck Laloë, Garrett Lisi (PSI), Burkard Polster (Mathologer), Destin Sandlin (Smarter Every Day), Dianna Cowern (Physics Girl), Grigori Volovik, Robert Brady, Ross Anderson, Erwin Madelung, Tim Maudlin, Richard Feynman, Craig Callendar, Christian Wüthrich, Cohl Furey, Yoshio Koide, David Richeson, William Thomson, TED, and The Royal Institution—for their gifts of clarity.

I thank Johnny Wanda, Nicholas Cecaci, Avi Rubin, and David Heggli for their editorial feedback and Pashaa Sanwick for creating the cover.

And finally, I thank Avi Ruben and David Heggli for supporting me in this journey.

Other books by Thad:

Einstein's Intuition: Visualizing Nature in Eleven Dimensions

Moon Rock: Mare Crisium

Passages

A Perfect Universe

44

www.ingramcontent.com/pod-product-compliance
Lightning Source LLC
Chambersburg PA
CBHW041146210326
41519CB00046B/141
* 9 7 8 0 9 9 6 3 9 4 2 6 0 *